Integration of AC/DC Microgrids into Power Grids

Integration of AC/DC Microgrids into Power Grids

Special Issue Editor

Fazel Mohammadi

MDPI • Basel • Beijing • Wuhan • Barcelona • Belgrade • Manchester • Tokyo • Cluj • Tianjin

Special Issue Editor
Fazel Mohammadi
University of Windsor
Canada

Editorial Office
MDPI
St. Alban-Anlage 66
4052 Basel, Switzerland

This is a reprint of articles from the Special Issue published online in the open access journal *Sustainability* (ISSN 2071-1050) (available at: https://www.mdpi.com/journal/sustainability/special_issues/ACDC_Microgrids).

For citation purposes, cite each article independently as indicated on the article page online and as indicated below:

LastName, A.A.; LastName, B.B.; LastName, C.C. Article Title. *Journal Name* **Year**, *Article Number*, Page Range.

ISBN 978-3-03936-180-9 (Pbk)
ISBN 978-3-03936-181-6 (PDF)

Contents

About the Special Issue Editor

Fazel Mohammadi (Ph.D., SM' IEEE) is the founder of the Power and Energy Systems Research Laboratory. He is a Senior Member of Institute of Electrical and Electronics Engineers (IEEE), and an active member of International Council on Large Electric Systems (CIGRE), European Power Electronics and Drives (EPE) Association, American Wind Energy Association (AWEA), and the Institution of Engineering and Technology (IET). His research interests include Power and Energy Systems Control, Operation, Planning, and Reliability, High-Voltage Engineering, Power Electronics, and Smart Grid.

Editorial

Integration of AC/DC Microgrids into Power Grids

Fazel Mohammadi

Electrical and Computer Engineering (ECE) Department, University of Windsor, Windsor, ON N9B 1K3, Canada; fazel@uwindsor.ca or fazel.mohammadi@ieee.org

Received: 13 April 2020; Accepted: 16 April 2020; Published: 19 April 2020

Abstract: The Special Issue on "Integration of AC/DC Microgrids into Power Grids" is published. A total of six qualified papers are published in this Special Issue. The topics of the papers are the Optimal Power Flow (OPF), control, protection, and the operation of hybrid AC/DC microgrids. Nine researchers participated in this Special Issue. We hope that this Special Issue is helpful for sustainable energy applications.

Keywords: AC/DC Microgrids; Distributed Generations (DGs); Microgrid Control Systems; Power Systems Operation; Power Systems Optimization; Power Systems Planning; Power Systems Protection

---

1. Introduction

AC/DC microgrids are a small part of low voltage distribution networks that are placed far from power substations and also are also interconnected through the Point of Common Coupling (PCC) to power grids [1–3]. These systems are the important keys to use flexible, techno-economic, and environmentally-friendly generation units for the reliable operation and cost-effective planning of smart electricity grids [4,5]. Although AC/DC microgrids with the integration of renewable energy resources and other energy systems, such as power-to-gas, combined heat and power, combined cooling heat and power, power-to-heat, power-to-vehicle, and pump and compressed air storage, have several advantages, there are some technical aspects that must be addressed. This Special Issue aims to study the configuration, impacts, and prospects of AC/DC microgrids that enable enhanced solutions for intelligent and optimized electricity systems, energy storage systems, and demand-side management in power grids with an increasing share of distributed energy resources. This includes AC/DC microgrids modeling, simulation, control, operation, protection and dynamics along with considering power quality improvement, load forecasting, energy conversion, supervisory and monitoring, diagnostics and prognostics systems.

2. Integration of AC/DC Microgrids into Power Systems

T. V. Nguyen, et al. [6], in their paper entitled "Power Flow Control Strategy and Reliable DC-Link Voltage Restoration for DC Microgrid under Grid Fault Conditions", proposes an effective Power Flow Control Strategy (PFCS) based on the centralized control method and a reliable DC-link Voltage (DCV) restoration algorithm for a DC microgrid, which consists of a utility grid connection system, a wind power generation system, a battery-based energy storage system, and DC loads, under grid fault conditions. The power flow in a DC microgrid can be autonomously and reliably controlled under both the grid-connected and islanded conditions using the relationship of supply demand power and battery status. By implementing the Constant Current-Constant Voltage (CC-CV) method for the battery-charging operation with the consideration of the battery power limit, the overheating or damage caused by undesirable over-charging/over-discharging is avoided, which expands the battery life, significantly. This paper also develops an effective load shedding algorithm considering the State-of-Charge (SoC) and the maximum capability of the battery to maintain the system power balance, even in the critical cases. In order to deal with the system power imbalance caused by the delay of grid fault detection, the DCV restoration algorithm is proposed in this paper. In the proposed DCV restoration algorithm, a Local Emergency Control Mode (LECM) is introduced to restore the DCV quickly to a nominal value. The LECM, which is achieved by local controllers either with the battery-based energy storage system or the wind power generation system, operates regardless of the control signals from the central controller under critical conditions for the purpose of ensuring the system power balance. To validate the effectiveness of the PFCS and the proposed DCV restoration algorithm, both simulations based on the PSIM software and experiments based on the prototype laboratory DC microgrid testbed are carried out.

F. Mohammadi, et al. [7], in their paper entitled "A Bidirectional Power Charging Control Strategy for Plug-in Hybrid Electric Vehicles", proposes a precise bidirectional charging control strategy of Plug-in Hybrid Electric Vehicles (PHEVs) in power grids to simultaneously regulate the voltage and frequency, reduce the peak load, and improve the power quality considering the SoC and available active power in power grids. Different events that may occur during a 24-h scenario in the studied Distributed Generation (DG)-based system consisting of different microgrids, diesel generator and wind farm, PHEVs with several charging profiles, and different loads, are considered. The simulation and analysis are performed in MATLAB/SIMULINK software. The simulation results show the robustness of the proposed control strategy.

F. Mohammadi, et al. [8], in their paper entitled "A New Topology of a Fast Proactive Hybrid DC Circuit Breaker for MT-HVDC Grids", proposes a new topology of a fast proactive Hybrid DC Breaker (HDCCB) to isolate the DC faults in Multi-Terminal VSC-HVDC (MT-HVDC) grids in case of fault current interruption, along with lowering the conduction losses and lowering the interruption time. All modes of operation of the proposed topology are studied. The simulation results verify that compared to the conventional DC circuit breaker, the proposed topology has a lower interruption time, lower on-state switching losses, and a higher breaking current capability. Due to the fact that MT-HVDC systems can share the power among the converter stations independently and in both directions, and considering the fast bidirectional fault current interruption, as well as reclosing and rebreaking capabilities, the proposed HDCCB can improve the overall performance of MT-HVDC systems and increase the reliability of the DC grids.

M. A. Hassan, et al. [9], in their paper entitled "Optimal Power Control of Inverter-Based Distributed Generations in Grid-Connected Microgrid", proposes an efficient PI power controller to regulate the predefined injected real and reactive powers to the grid. The control problem is optimally designed based on minimizing the error between the calculated and the injected powers to get the optimal controller parameters. Particle Swarm Optimization (PSO) is employed to design the controller parameters and LC filter components. Different microgrid structures are implemented and examined in MATLAB. Firstly, the optimal proposed controller is designed to control the injected real and reactive powers of one inverter-based DG. Secondly, the optimal proposed controller is designed for two different rated inverter-based DG units to share their injected powers to the grid. Several disturbances, such as step-up/down changes of real and reactive injected powers, three-phase faults, and losing DG units are applied to investigate the effectiveness of the proposed controller and to ensure the system stability after getting disturbed. Additionally, to validate the usefulness of the proposed controller, the considered microgrid is implemented on a Real-Time Digital Simulator (RTDS). To confirm the effectiveness of the proposed optimal control scheme, it is compared with the existing work through extensive simulation and experiments under various disturbances. The results confirm the superiority of the proposed control strategy in providing a fast, accurate and decoupled power control with a lower AC current distortion.

T. V. Nguyen, et al. [10], in their paper entitled "An Improved Power Management Strategy for MAS-Based Distributed Control of DC Microgrid under Communication Network Problems", proposes an improved Power Management Strategy (PMS) using the Multi-Agent System (MAS)-based distributed control of the DC microgrid. In this study, the DC microgrid consists of a grid agent, a battery agent, a wind power generation system agent, and a load agent. To ensure the system power balance under various conditions, each agent investigates the information obtained from both the local measurement and the neighboring agents via the communication lines. Then, the decision for local control mode and communication data is optimally made for the system power balance. By using this control scheme, the control mode of agents could be determined locally without any intervention of the central controller, which effectively avoids the single point of failure as in the centralized control. Moreover, all the agents can operate in a deliberative and cooperative manner to ensure globally the optimal operation by the means of the communication network. In addition, to deal with the impact of communication problems in the case of the grid fault, a DCV restoration algorithm is introduced to restore the DCV stably to its nominal value. Furthermore, to recognize the grid recovery reliably in other agents even under communication failure, the grid recovery identification algorithm is introduced. For this purpose, a special current pattern is generated on the DC-link by the grid agent once the grid is recovered. By detecting this current pattern on the DC-link, the remaining agents can reliably identify the grid recovery even without the communication. To validate the feasibility of the MAS-based distributed control as well as the proposed schemes under the communication problems, the simulations based on the PSIM software and the experiments based on laboratory prototype DC microgrid testbed are carried out.

F. Mohammadi, et al. [11], in their paper entitled "Optimal Placement of TCSC for Congestion Management and Power Loss Reduction Using Multi-Objective Genetic Algorithm", proposes a multi-objective algorithm to optimally allocate Thyristor-Controlled Series Compensators (TCSCs) and their susceptance values, considering power loss reduction, congestion management, and the determination of the power lines compensation rates in power grids. This paper focuses on (1) considering the structure of TCSCs and the AC characteristics of power systems, formulating a nonlinear problem, and solving it using a heuristic algorithm, and (2) determining the optimal allocation and rating of TCSCs through an optimization procedure. The Jacobian sensitivity approach and AC load flow are used for line congestion evaluations. The Multi-Objective Genetic Algorithm (MOGA) is used as an optimization method to determine the optimal locations and susceptance values of TCSCs. The proposed method is deployed on the IEEE 30-bus test system, and the results are investigated to illustrate the applicability and effectiveness of the proposed method. In addition, the obtained results are compared with those from different algorithms, such as the multi-objective PSO algorithm, Differential Evolution (DE) algorithm, and the Mixed-Integer Non-Linear Program (MINLP) technique. The obtained results show the superiorities of the proposed method, in terms of fast convergence and high accuracy, over the other heuristic methods.

Funding: This research received no external funding.

Acknowledgments: I am thankful for the contributions of professional authors and reviewers and the excellent assistant of the editorial team of Sustainability.

Conflicts of Interest: The author declares no conflict of interest.

References

1. Mohammadi, F.; Nazri, G.-A.; Saif, M. An Improved Droop-Based Control Strategy for MT-HVDC Systems. *Electronics* **2020**, *9*, 87. [CrossRef]
2. Mohammadi, F.; Nazri, G.-A.; Saif, M. An Improved Mixed AC/DC Power Flow Algorithm in Hybrid AC/DC Grids with MT-HVDC Systems. *Appl. Sci.* **2020**, *10*, 297. [CrossRef]
3. Mohammadi, F. Power Management Strategy in Multi-Terminal VSC-HVDC System. In Proceedings of the 4th National Conference on Applied Research in Electrical, Mechanical Computer and IT Engineering, Tehran, Iran, 4 October 2018.
4. Mohammadi, F.; Zheng, C. Stability Analysis of Electric Power System. In Proceedings of the 4th National Conference on Technology in Electrical and Computer Engineering, Tehran, Iran, 27 December 2018.
5. Mohammadi, F.; Nazri, G.A.; Saif, M. A Fast Fault Detection and Identification Approach in Power Distribution Systems. In Proceedings of the IEEE 5th International Conference on Power Generation Systems and Renewable Energy Technologies (PGSRET), Istanbul, Turkey, 26–27 August 2019.

6. Van Nguyen, T.; Kim, K.-H. Power Flow Control Strategy and Reliable DC-Link Voltage Restoration for DC Microgrid under Grid Fault Conditions. *Sustainability* **2019**, *11*, 3781. [CrossRef]
7. Mohammadi, F.; Nazri, G.-A.; Saif, M. A Bidirectional Power Charging Control Strategy for Plug-in Hybrid Electric Vehicles. *Sustainability* **2019**, *11*, 4317. [CrossRef]
8. Mohammadi, F.; Nazri, G.-A.; Saif, M. A New Topology of a Fast Proactive Hybrid DC Circuit Breaker for MT-HVDC Grids. *Sustainability* **2019**, *11*, 4493. [CrossRef]
9. Hassan, M.A.; Worku, M.Y.; Abido, M.A. Optimal Power Control of Inverter-Based Distributed Generations in Grid-Connected Microgrid. *Sustainability* **2019**, *11*, 5828. [CrossRef]
10. Nguyen, T.V.; Kim, K.-H. An Improved Power Management Strategy for MAS-Based Distributed Control of DC Microgrid under Communication Network Problems. *Sustainability* **2020**, *12*, 122. [CrossRef]
11. Nguyen, T.T.; Mohammadi, F. Optimal Placement of TCSC for Congestion Management and Power Loss Reduction Using Multi-Objective Genetic Algorithm. *Sustainability* **2020**, *12*, 2813. [CrossRef]

Article

Power Flow Control Strategy and Reliable DC-Link Voltage Restoration for DC Microgrid under Grid Fault Conditions

Thanh Van Nguyen and Kyeong-Hwa Kim *

Department of Electrical and Information Engineering, Seoul National University of Science and Technology, 232 Gongneung-ro, Nowon-gu, Seoul 01811, Korea
* Correspondence: k2h1@seoultech.ac.kr; Tel.: +82-2-970-6406

Received: 1 June 2019; Accepted: 6 July 2019; Published: 10 July 2019

Abstract: In this paper, an effective power flow control strategy (PFCS) based on the centralized control method and a reliable DC-link voltage (DCV) restoration algorithm for a DC microgrid (DCMG) under grid fault conditions are proposed. Considering the relationship of supply-demand power and the statuses of system units, thirteen operating modes are presented to ensure the power balance in DCMG under various conditions. In the PFCS, the battery charging/discharging procedure is implemented considering the battery power limit to avoid overheating and damage. Moreover, load shedding and load reconnection algorithms are presented to maintain the system power balance, even in critical cases. To prevent the system power imbalance in DCMG caused by the delay of grid fault detection, a reliable DCV restoration algorithm is also proposed in this paper. In the proposed scheme, as soon as abnormal behavior of the DCV is detected, the battery or wind power generation system instantly enters a local emergency control mode to restore the DCV rapidly to the nominal value, regardless of the control mode assigned from the central controller. Comprehensive simulations and experiments based on the DCMG testbed are carried out to prove the effectiveness of the PFCS and the proposed DCV restoration algorithm.

Keywords: DC microgrid; experimental evaluation; grid fault conditions; power flow control strategy; reliable DC-link voltage restoration

---

1. Introduction

In recent years, renewable energy sources (RESs) such as wind and solar are attracting a great deal of attention due to the scarcity of fossil energy and environmental issues [1]. For the purpose of integrating various RESs into the utility grid (UG), distributed generations (DGs), which are the main parts of microgrids (MGs), have been considered and developed [2]. Due to the advantages of easy resource integration, flexible installation location, and reliable operation, DG-based MGs have become a future trend in constructing electric power systems [3]. Energy storage systems (ESSs) are usually integrated with MG in order to improve the reliability of system operation by reducing the effect of the intermittent nature of RESs [4].

According to the types of bus voltage, MGs can be classified into DC microgrids (DCMGs), AC microgrids (ACMGs), and hybrid AC/DC microgrids [5]. In comparison with other configurations, the control of DCMG is simpler, since the system power balance can be guaranteed only by regulating the DC-link voltage (DCV) to its nominal value without consideration of reactive power, current harmonics, frequency stability, and phase imbalance [6,7].

Regarding the communication perspective, the DCMG control method can be mainly divided into three categories: decentralized control [8–10], distributed control [11,12], and centralized control [6,13,14]. In decentralized control, DC-bus signaling (DBS) is well known as a simple and effective autonomous power control method [9]. In the DBS approach, the DCV is used as an indicator to determine the operating mode of system units according to predefined voltage thresholds. Even though the DBS method has the advantage of simplicity of control, its performance depends significantly on the selection of appropriate voltage thresholds. If the difference between two voltage thresholds is too large, the DCV begins to fluctuate. In contrast, if the selected difference is too small, the determination of operating modes is strongly affected by sensor inaccuracies, which may lead to unnecessary and frequent mode switching [15]. In Ref [10], a communication-free method is proposed to manage the power in the system. In this study, the power management strategy is based on the fixed setpoints of the DCV and the active power outputs. Generally, the decentralized control has the performance limitations of the entire system control, since each system unit lacks information about the others. In order to overcome this weakness, the distributed control, which constructs a simple communication network to share the information among neighboring system units, has been considered and developed [11]. Because only neighboring units are connected, a consensus algorithm is required in this method to obtain global information. In spite of the advantages of distributed control, it still has the limitations, such as the effect of measurement errors and the complexity of analytical performance analyses for the convergence speed of the consensus algorithm [12,15]. Unlike the above two methods, in the centralized control approach, a central controller (CC) is formed to collect the necessary data from local controllers (LCs), which then sends optimal decisions back to LCs to control power exchanges and to guarantee the system's stability under various conditions [5]. By implementing CC, other concerns in DCMG, such as the cost minimization [13] and adaptive protection [14], can also be realized easily.

Recently, many power flow control strategies (PFCSs) based on the centralized control approach have been presented for the purpose of ensuring the system power balance under various conditions [16–19]. In Ref [16], an improved voltage control strategy for a DCMG composed of photovoltaic and hybrid ESSs, such as batteries and supercapacitors, is proposed to stabilize the DCV. In this scheme, the entire system operation is divided into four states with individual control modes considering the photovoltaic generation power, load demand, and battery state of charge (*SOC*). In Ref [17], a PFCS based on the peak load demand durations in UG, and the optimal usage of RES is studied. However, in this work, the maximum charging and discharging power limit of the battery are not taken into account, which may affect the power control strategy in some cases. For instance, in case of power shortage, it is probable that the DCMG would request the discharging power from battery, exceeding the maximum capability. Similarly, the required charging power to the battery may be greater than the maximum absorbed capability in power surplus situations. In both cases, maintaining the power balance in DCMG by a battery is not possible due to the power limit. To deal with this issue, the maximum charging and discharging currents of the battery are considered in the design of PFCS under the grid-connected mode in Ref [18]. In Ref [19,20], to avoid the system collapse in case of power shortage, an appropriate load shedding (LS) is implemented to disconnect less important loads. On the other hand, to prevent the DC-link from overvoltage in case of power surplus, another approach halts the RESs conversion system [9] or uses the voltage control mode (VCM) [16,21]. In practice, it is necessary to reconnect some disconnected loads as the entire system becomes stable after the LS action [22]. Even though a load reconnection (LR) is considered in Ref [23,24], a reliable sequence of reconnection is neither completely analyzed nor presented in these studies.

In the centralized control approach, the UG is a main source to guarantee the system power balance in the grid-connected mode. When a fault in the UG is detected by a fault detection device, the fault signal is sent to the CC by a communication line. Then, the CC changes the system operation into the islanded mode and assigns other units such as battery or wind power generation systems (WPGS) to maintain the system power balance. In practice, however, the CC cannot instantly change the system operation into the islanded mode due to the delay caused by fault clearance time, data transmission time, and the processing time of CC. Consequently, any source or power converter does not take on the role of temporarily regulating the DCV, resulting in a power system imbalance. To address this problem, much research focusing on grid fault detection schemes has been carried out for the purpose of reducing the time delay [25,26]. However, detecting the grid fault accurately and rapidly is still a challenging issue [27].

This paper presents an effective PFCS and a reliable DCV restoration algorithm for DCMG under grid fault conditions. The effective PFCS is achieved based on the centralized control method for DCMG, which consists of a UG connection system, a WPGS, a battery-based ESS, and DC loads. The power flow in DCMG can be autonomously and reliably controlled under both the grid-connected and islanded conditions by using the relationship of supply-demand power and battery status. By implementing the constant current-constant voltage (CC-CV) method for the battery charging operation with the consideration of the battery power limit, the overheating or damage caused by undesirable overcharging/overdischarging is avoided, which expands the battery life significantly. This paper also develops an effective LS algorithm considering the *SOC* and the maximum capability of the battery to maintain the system power balance, even in the critical cases. Beside LS, an LR algorithm is also implemented for the purpose of reconnecting the loads which are initially disconnected due to power shortage in DCMG. In both the LS and LR algorithms, a time delay is applied to avoid undesirable load disconnections or reconnections caused by noise. In order to deal with the system power imbalance caused by the delay of grid fault detection, the DCV restoration algorithm is proposed in this paper. In the proposed DCV restoration algorithm, a local emergency control mode (LECM) is introduced to restore the DCV quickly to a nominal value. The LECM, which is achieved by LCs either with the battery-based ESS or the WPGS, operates regardless of the control signals from the CC under critical conditions for the purpose of ensuring the system power balance. To validate the effectiveness of the PFCS including LS and LR, as well as the proposed DCV restoration algorithm, both simulations based on the PSIM software and experiments based on prototype laboratory DCMG testbed are carried out.

This paper is organized as follows: Section 2 describes the configuration of DCMG. The details of the PFCS and system control methods of DCMG are discussed in Section 3. Section 4 presents the proposed DCV restoration algorithm. The simulation and experimental results are shown in Sections 5 and 6, respectively. Section 7 presents the discussion. Finally, Section 8 concludes the paper.

2. System Configuration of DCMG

Figure 1 shows the configuration of a DCMG which consists of four main units, namely, WPGS, battery-based ESS, UG connection system, and DC loads, where P_G is output power of UG connection system, P_B is battery power, P_W is output power of WPGS, and P_L is load power. In the WPGS, a permanent magnet synchronous generator (PMSG) is employed to convert the wind turbine mechanical output power into electrical power. The PMSG is widely used in WPGS because it has a simple structure, high efficiency, and a wide operating range. The output power of the PMSG is injected into the DC-link through a unidirectional AC/DC converter. Due to the fluctuation of wind power, battery-based ESS is normally coordinated with WPGS in DCMG to stabilize the DCV and system power flow. To interface the battery-based ESS, an interleaved bidirectional DC/DC converter, which exchanges the power between the battery and DC-link, is employed. By using the interleaved bidirectional converter, the output current ripples are notably reduced. As a result, this improves the charging performance and extends the battery life [28]. For the purpose of interacting DCMG with UG in the grid-connected mode, a UG connection system, including a transformer and a bidirectional AC/DC converter, is constructed. The UG connection system not only maintains a supply-demand power balance of the DCMG, but also injects high quality current into UG depending on the DCMG operation modes. In the system configuration, a CC determines all the operating modes of the DCMG units. A power converter equipped with LC is implemented to operate each DCMG unit according to the control mode assigned from the CC. Load management is also achieved by the CC to accomplish the LS and LR algorithms.

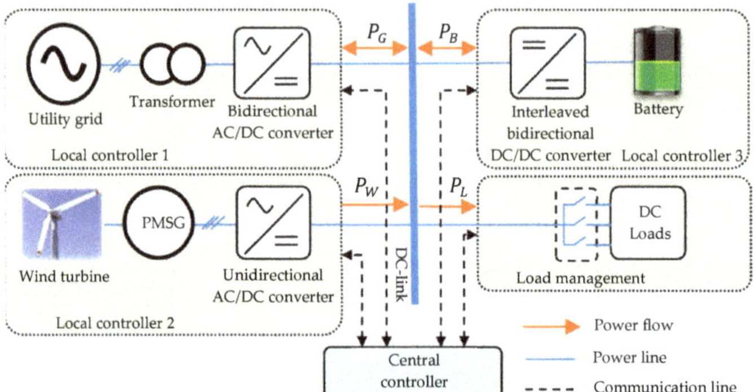

Figure 1. Configuration of DCMG.

3. Power Flow Control Strategy and System Control of DCMG

3.1. Power Flow Control Strategy

Figure 2 shows the PFCS of DCMG in this study. Based on the correlation of wind power and load demand, UG status, and battery status, thirteen operating modes of DCMG are determined to ensure the system power balance under various conditions. The control and operation of DCMG units according to each operating mode are described in Table 1 in detail. The control mode with an asterisk in the power converter indicates that this converter is in charge of the DCV regulation and system power balance. Symbols in Figure 2 and Table 1 are defined as follows: SOC^{max} is the maximum battery state of charge; SOC^{min} is the minimum battery state of charge; $I^{req}_{B,cha}$ is the required battery charging current; $I^{req}_{B,dis}$ is the required battery discharging current; $I^{max}_{B,cha}$ is the maximum battery charging current; $I^{max}_{B,dis}$ is the maximum battery discharging current; V_B is the battery voltage; V^{max}_B is the maximum battery voltage; REC denotes rectifier mode; INV denotes inverter mode; DIS denotes disconnected mode in case of UG fault; IDLE denotes idle mode; BVCM denotes battery voltage control mode; BCCM denotes battery current control mode; DCVM-D denotes DCV control mode by battery discharging; DCVM-C denotes DCV control mode by battery charging; SHED denotes LS mode; NC/RECO denotes no change/LR mode; MPPT denotes maximum power point tracking mode.

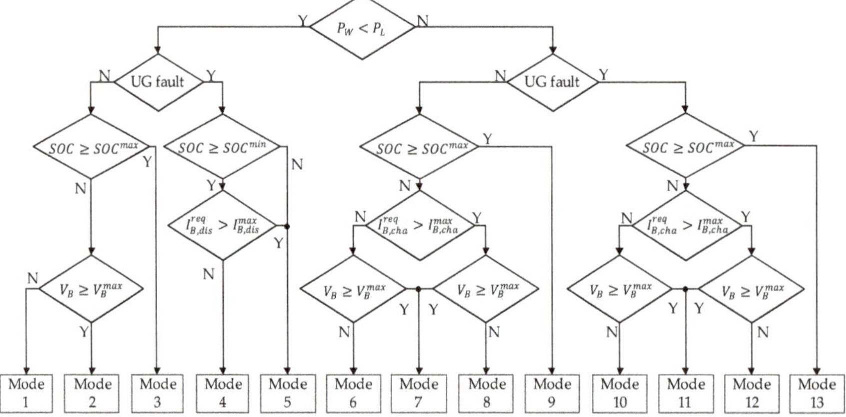

Figure 2. PFCS of DCMG.

Table 1. Detailed description of thirteen operating modes in PFCS.

Mode	WPGS	Load	UG Connection System	Battery-Based ESS
Mode 1	MPPT	NC/RECO	REC*	BCCM
Mode 2	MPPT	NC/RECO	REC*	BVCM
Mode 3	MPPT	NC/RECO	REC*	IDLE
Mode 4	MPPT	NC/RECO	DIS	DCVM-D*
Mode 5	MPPT	SHED	DIS	IDLE/DCVM-D*
Mode 6	MPPT	NC/RECO	IDLE	DCVM-C*
Mode 7	MPPT	NC/RECO	INV*	BVCM
Mode 8	MPPT	NC/RECO	INV*	BCCM
Mode 9	MPPT	NC/RECO	INV*	IDLE
Mode 10	MPPT	NC/RECO	DIS	DCVM-C*
Mode 11	VCM*	NC/RECO	DIS	BVCM
Mode 12	VCM*	NC/RECO	DIS	BCCM
Mode 13	VCM*	NC/RECO	DIS	IDLE

The entire control and operation of DCMG units corresponding to the thirteen operating modes are explained as follows.

Operating mode 1: The WPGS is operated in the MPPT mode to inject the maximum power from the wind turbine into the DC-link. In this mode, however, the injected power to the DC-link is not sufficient to supply the load demand ($P_W < P_L$). In this case, the power deficit is compensated by the UG via REC mode operation of AC/DC bidirectional converter. All loads are fed in this operating mode. The battery is controlled in BCCM with the charging current of $C/10$, where C denotes the rated capacity of battery [29]. This charging current level is selected to prevent the battery from overheating and damage.

Operating mode 2: Similar to operating mode 1, the WPGS is operated in the MPPT mode. Because wind power is still not sufficient to supply the load, the DCV is regulated by the UG through AC/DC bidirectional converter with REC mode operation. The only difference from operating mode 1 is that the battery is operated in BVCM in order to avoid the battery damage caused by overvoltage since the battery voltage is greater than its maximum value of V_B^{max}. All loads are fed in this operating mode.

Operating mode 3: The battery SOC reaches its maximum value of SOC^{max}. To prevent the battery from overcharging, the battery operation is switched into IDLE mode. The WPGS is still operated in the MPPT mode and the entire power balance is maintained by the UG. All loads are fed in this operating mode.

Operating mode 4: When a grid fault occurs, the UG is disconnected from DCMG. This mode corresponds to the islanded operating mode of the DCMG. Since the battery SOC is higher than the minimum value of SOC^{min} and the required battery discharging current is below the allowable maximum limitation ($I_{B,dis}^{req} \leq I_{B,dis}^{max}$), the battery starts the discharging in DCVM-D mode to regulate the DCV. The WPGS works in the MPPT mode and all loads are fed.

Operating mode 5: This operating mode occurs when the required battery discharging current exceeds its maximum rating ($I_{B,dis}^{req} > I_{B,dis}^{max}$) or the battery SOC is lower than the minimum level ($SOC < SOC^{min}$). In order to prevent the battery from overdischarging, in case of $I_{B,dis}^{req} > I_{B,dis}^{max}$, the battery is discharged with the maximum discharging current, which is achieved by a current limiter in DCVM-D mode. In the latter condition, the battery operation is switched into IDLE mode. In both cases, the battery is not able to regulate the DCV any longer. To avoid system collapse under this critical condition, the LS is activated to disconnect some less important loads. After the LS is completed, there are three possible cases that may happen next as seen from Figure 2. If $I_{B,dis}^{req}$ becomes smaller than $I_{B,dis}^{max}$ after the LS, the system operation is returned to operating mode 4, in which the battery undertakes the task of the DCV control again by means of DCVM-D mode. In another situation, if the remaining load demand becomes lower than the power extracted from the WPGS, and the required battery charging current $I_{B,cha}^{req}$ is smaller than $I_{B,cha}^{max}$, the system operation is switched into operating mode 10, in which the battery regulates the DCV by DCVM-C mode. Otherwise, the system operation is switched to operating mode 12, in which the WPGS starts to control the DCV by VCM in case that the required battery charging current $I_{B,cha}^{req}$ is greater than $I_{B,cha}^{max}$. From the above analysis, it is worth mentioning that operating mode 5 only works as a transient mode. After the LS is accomplished at this mode, the system operation is changed to other operating modes, in which the DCV is continuously regulated by the battery or WPGS.

Operating mode 6: This operating mode corresponds to the system operation when the wind power is higher than load demand. If the available power on the DC-link is large enough and the system is stable, the disconnected load in operating mode 5 can be reconnected by using LR algorithm. Since the battery *SOC* is smaller than the maximum value of SOC^{max}, and the battery voltage and required charging current are less than their predefined rating ($V_B < V_B^{max}$ and $I_{B,cha}^{req} \leq I_{B,cha}^{max}$), the battery has a capability to regulate the DCV by using DCVM-C mode. Since the UG is available in this case, the UG may control the DCV through INV mode. However, by considering the unpredictable fault of the UG as well as the variation of wind power, the battery operation with DCVM-C is chosen by priority, which ensures that the battery power can be maintained as high as possible to deal with later critical conditions. As a result, the UG connection system enters IDLE mode, which indicates the UG does not participate in the power exchange of system.

Operating mode 7: Similar to operating mode 6, the wind power is higher than load demand without the UG fault. However, the battery voltage reaches the maximum value of V_B^{max}. To protect the battery from overvoltage, the battery is charged with BVCM mode, absorbing a portion of the surplus power. The remaining surplus power is absorbed by the UG through INV mode of the bidirectional AC/DC converter.

Operating mode 8: In this mode, the wind power is higher than the load demand without the UG fault. In addition, the required charging current is higher than the maximum rating ($I_{B,cha}^{req} > I_{B,cha}^{max}$). In this case, the battery is controlled by charging it with the maximum charging current of $I_{B,cha}^{max}$. Similar to operating mode 7, the UG takes charge of the DCV regulation through INV mode of bidirectional AC/DC converter.

Operating mode 9: As the battery *SOC* reaches the maximum value of SOC^{max}, the battery operation is changed into IDLE mode to prevent the battery from overcharging. The UG conducts the role of the DCV regulation by INV mode.

Operating mode 10: This mode corresponds to the islanded operation due to grid fault. In this condition, the battery *SOC* is smaller than the maximum SOC^{max}, also the required charging current and the battery voltage are still less than their maximum rating ($V_B < V_B^{max}$ and $I_{B,cha}^{req} \leq I_{B,cha}^{max}$). In this case, the battery can effectively regulate the DCV in DCVM-C mode by absorbing surplus power.

Operating mode 11: As the battery voltage reaches the maximum value of V_B^{max}, the battery operation is changed from DCVM-C to BVCM to avoid battery overvoltage. As a result, the authority to control the DCV is released to the power converter of WPGS. For this purpose, the WPGS changes the control mode into VCM in which the DCV is maintained at the nominal value by limiting the injected power from wind to DC-link.

Operating mode 12: This operating mode is the same as the operating mode 8 except for the UG fault. The required battery charging current is higher than the maximum rating ($I_{B,cha}^{req} > I_{B,cha}^{max}$) due to the continuous increase in wind power. As a result, the battery is incapable of regulating the DCV. In this case, BCCM is applied to charge the battery with the maximum charging current and the DCV is kept stable by the WPGS with VCM.

Operating mode 13: When the battery *SOC* reaches the maximum value of SOC^{max}, the battery operation is changed into IDLE mode same with operating mode 9 to avoid overcharging. The WPGS is still in charge of the DCV regulation with VCM.

3.2. Control Scheme of Gird-Connected Converter

Figure 3 shows the control loops of the bidirectional AC/DC converter (converter 1) used for the UG connection, where V_{DC} is the DCV, I_d and I_q are currents in the synchronous reference frame (SRF), and superscript '*ref*' denotes the reference quantity. As seen in Figure 3, the DCV is controlled by an outer loop proportional-integral (PI) controller either in REC or in INV mode. The outer loop DCV controller generates a positive reference current in INV mode, and a negative one in REC mode. The limiters are located after the PI controllers to avoid high undesirable reference currents. A decoupling current controller in the SRF is used for the inner current control loop to ensure zero steady-state tracking error [10,30]. The obtained reference voltages are applied by a space vector modulation (SVM). The IDLE and DIS modes are easily implemented by switching off the connection switch between the UG and DCMG.

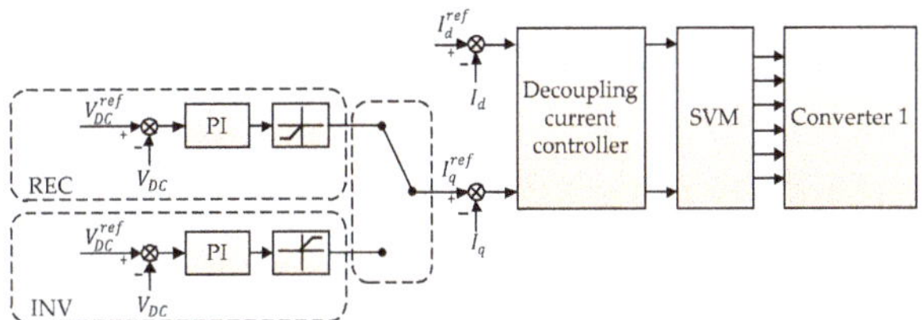

Figure 3. Control loops of converter 1 used for UG connection.

3.3. Control Scheme of Wind Power Converter

As described in Section 3.1, the WPGS is operated in two operating modes, namely, the MPPT and VCM modes. The MPPT mode is employed to maximize the power extracted from wind while the VCM mode is employed to regulate the DCV in DCMG. Figure 4 shows two control loops of unidirectional AC/DC converter (converter 2) for the WPGS.

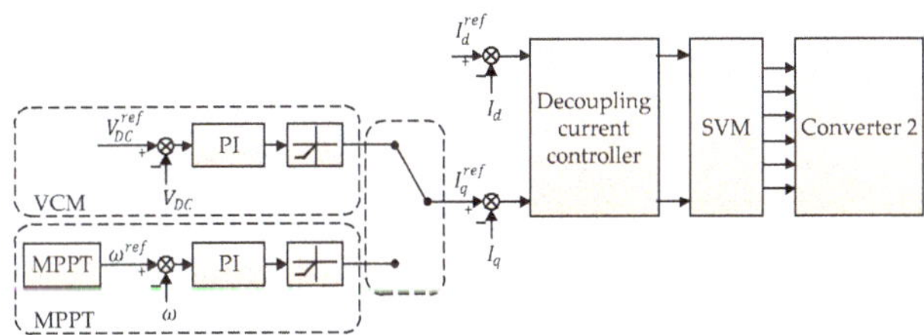

Figure 4. Control loops of converter 2 for WPGS.

The MPPT control algorithm is realized by three cascaded control loops. The outer loop determines the maximum power point using the MPPT algorithm. The output of the MPPT algorithm is the optimal rotor angular speed which is used as the reference of the inner control loop. Inner control loops are composed of two cascaded controllers. The first inner loop controls the rotor angular speed to the reference by using the PI controller. The control output generates the current references which are regulated by subsequent inner control loop based on the synchronous PI decoupling current controller.

In VCM mode, the control loop is cascaded by two control loops. The outer loop is realized to regulate the DCV by the PI controller. The controller outputs are employed as the current references in the inner control loop based on the synchronous PI decoupling current controller.

3.4. Control Scheme of Battery Energy Storage System

As explained in Section 3.1, the battery-based ESS is operated at five different operating modes, namely, DCVM-C, DCVM-D, BVCM, BCCM, and IDLE mode. The IDLE mode is easily accomplished by opening the connection switch between the battery and DC-link. Figure 5 shows the control loops for four other operating modes of interleaved bidirectional DC/DC converter (converter 3) to interface the battery-based ESS, where I_B is the battery current and d is the duty cycle. In both DCVM-C and DCVM-D modes, the DCV control is carried out in the outer control loop. By using the current reference generated from the outer control loop, the inner control loop regulates the charging or discharging currents to guarantee zero tracking errors of the current. In the battery system, the charging current is denoted by a positive value while the discharging current by a negative one.

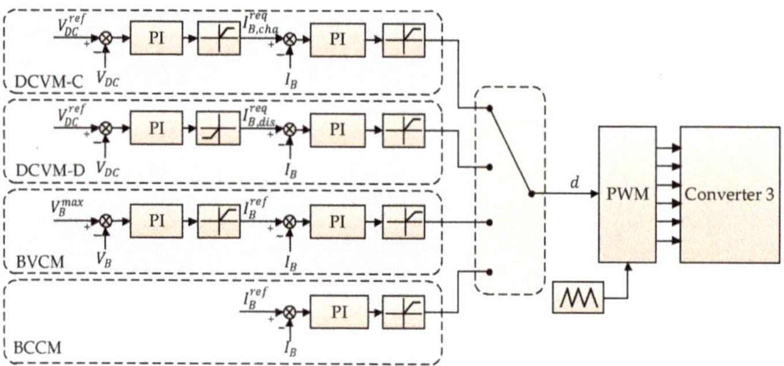

Figure 5. Control loops of converter 3 for battery-based ESS.

In BVCM, two control loops are employed to charge the battery with constant voltage. The outer control loop controls the battery at the maximum voltage (V_B^{max}) and the inner control loop is designed in the same way as the DCVM-C and DCVM-D modes. Unlike the above three modes, BCCM is realized by only one current control loop to charge the battery with a desired constant current. In all cases, the output of the current controller is the duty cycle d, which is fed to a pulse width modulation (PWM) block to drive the switches in DC/DC converter (converter 3).

3.5. Load Management Algorithm

Normally, when the battery is used as the main source to regulate the DCV in case of grid fault, the LS algorithm is not necessary. However, in some critical situations, the LS should be used as the last solution to prevent the system from collapsing. Figure 6 shows the LS algorithm presented in this paper, where i is flag describing the load status, n is the total quantity of load, C_{she}, T_{she}, and ΔT are the counter, predefined time delay, and step size, respectively.

As explained in operating mode 5, the LS is activated by one of two critical reasons that the required battery discharging current exceeds its maximum rating, or that the battery *SOC* is lower than the minimum level. The first critical case represents the power deficiency of battery, while the other comes from the insufficiency of energy stored in the battery. For the LS procedure, loads need to be classified into different priority levels to ensure that the load with lower priority should be disconnected first. In this paper, load priority levels are assigned as load 1 < load 2 < … < load n, where load 1 has the lowest priority and load n has the highest priority. Moreover, different shedding time delay levels are added into the LS algorithm to avoid undesirable load disconnection caused by noise. Table 2 shows the description of load status by using the flag i. During the LS, the algorithm checks the value of i to recognize the load which is already disconnected as well as disconnecting the load at next step. As shown in Figure 6, the counter C_{she} starts counting up with the step size of ΔT. As soon as C_{she} reaches the predefined T_{she}, the loads denoted by i are disconnected from DC-link according to Table 2. After load disconnection, C_{she} is reset for next operations.

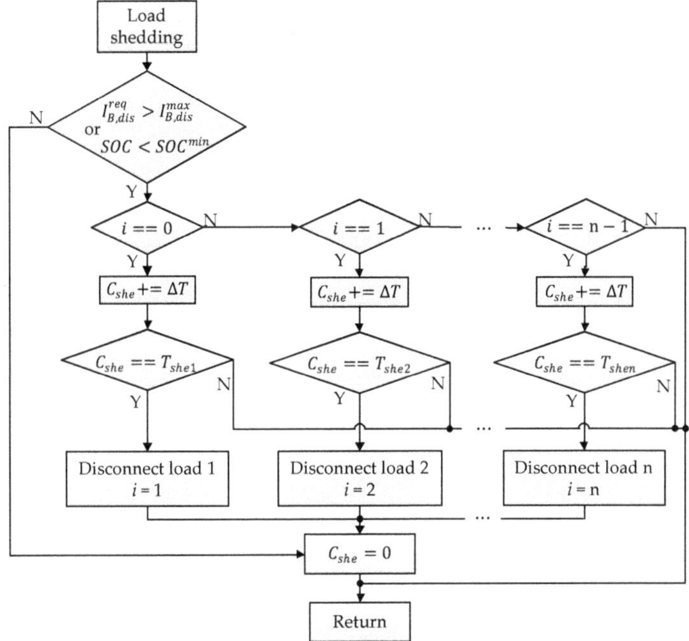

Figure 6. LS algorithm.

Table 2. Description of load status.

i	0	1	2	3	n
Indication	No load is disconnected	Load 1 (P_{L1}) is disconnected	Load 2, 1 (P_{L2}, P_{L1}) are disconnected	Load 3, 2, 1 (P_{L3}, P_{L2}, P_{L1}) are disconnected	All loads ($P_{Ln}, \ldots, P_{L2}, P_{L1}$) are disconnected

Once the system is back to normal, the disconnected load can be reconnected when necessary. In this paper, UG is assumed as a huge source which can feed any required consumed load. For instance, in case of grid connection, all the disconnected load can be reconnected regardless of the battery and WPGS status. In case of the islanded mode, however, the LR is determined based on the available power on DC-link which can be expressed as

$$P^{avail}_{DC} = P_W + P^{max}_{B,dis} - P_L \tag{1}$$

where P^{avail}_{DC} is the available power on DC-link and $P^{max}_{B,dis}$ is the maximum discharging power of the battery. Figure 7 shows the LR algorithm where C_{rec} and T_{rec} are the counter and predefined time delay, respectively. When the LR is activated, the algorithm checks the value of i to recognize the reconnecting load. If i is equal to zero, it indicates that no load has been disconnected. As a result, the reconnection algorithm is unnecessary. When i is not zero, the UG status is checked first to reconnect all the loads. In case of the UG fault, P^{avail}_{DC} is computed and is compared with the disconnected load having the highest priority to reconnect it first. The LR algorithm operates similarly to the LS algorithm, with a time delay of T_{rec} to avoid the effect due to noise. The counter C_{rec} starts counting and the LR is accomplished when C_{rec} reaches T_{rec}.

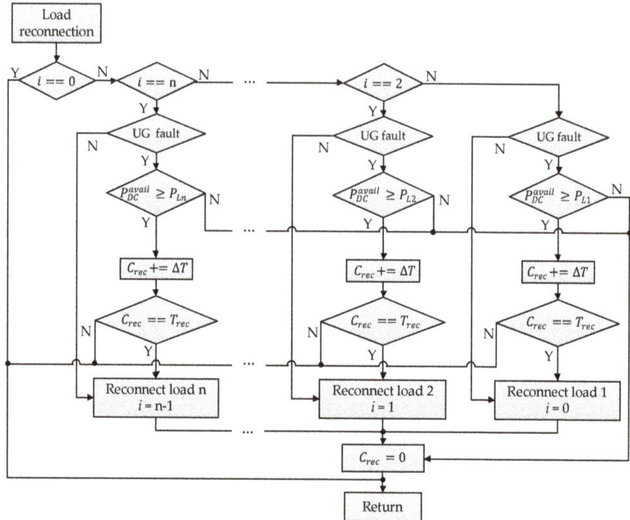

Figure 7. LR algorithm.

4. Proposed DC-Link Voltage Restoration Algorithm

In the grid-connected condition, converter 1 operates in the voltage control mode such as REC or INV to regulate the DCV. On the other hand, converter 2 works in the MPPT mode to maximize the extracted power from wind and converter 3 works in BCCM, BVCM, or IDLE to optimize the battery charging sequence. If the grid fault is detected right after fault occurrence, the system operation is quickly switched into the islanded mode in which the WPGS or battery undertakes the role of the DCV regulation. Unfortunately, the grid fault cannot be detected instantly by the fault detection algorithm in some cases due to a large total response time (TRT). In the centralized control approach, TRT is defined as the time interval from the fault occurrence to the final control action by the CC, which includes the fault clearance time, the data transmission time from the fault detection device to the CC, and the processing time for the controller to make a decision. Figure 8 shows TRT period when the grid fault occurs. In Figure 8, the fault clearance time, which consists of sensing time and opening time, is defined as the time duration from the fault occurrence instant to the instant at which DCMG is separated from UG [31]. Because the data transmission time and processing time are negligibly small in comparison with the fault clearance time, the delay of fault detection is considered to be equal to the fault clearance time. During the TRT period, any source or power converter does not serve to regulate the DCV. As a result, the DCV may be increased or decreased rapidly due to the imbalance between the supplied and consumed power, which ends up a catastrophic situation in DCMG. In order to prevent such a circumstance, an effective DCV restoration algorithm based on LECM is proposed in this paper. Figure 9 describes the control structure including the LECM for the proposed DCV restoration. It is worth noting that this paper focuses only on the effect of grid fault detection delays on system stability; the fault detection method is not considered.

In the normal DCMG operation, the CC collects the data from local systems, and then, assigns the proper execution mode (EM) to each LC and load connection/disconnection switches through communication lines as shown in Figure 9, in which EMj (for j = 1, 2, 3, 4) denotes the EM assigned to LCj (for j = 1, 2, 3) and load management system. During the normal operation, LCs are operated according to the EM assigned from the CC as specified in Table 1. For example, the WPGS has either MPPT or VCM as normal control mode as shown in Figure 9. In addition to these normal operation modes, LCs of both the WPGS and battery are also equipped with additional LECMs to deal with abnormal DCV variations caused by the delay of grid fault detection as shown in Figure 9, in which the WPGS has only VCM and the battery has either BCCM or DCVM-D as LECM. The proposed DCV restoration algorithm based on LECM for the WPGS and battery are presented in Figure 10 in detail.

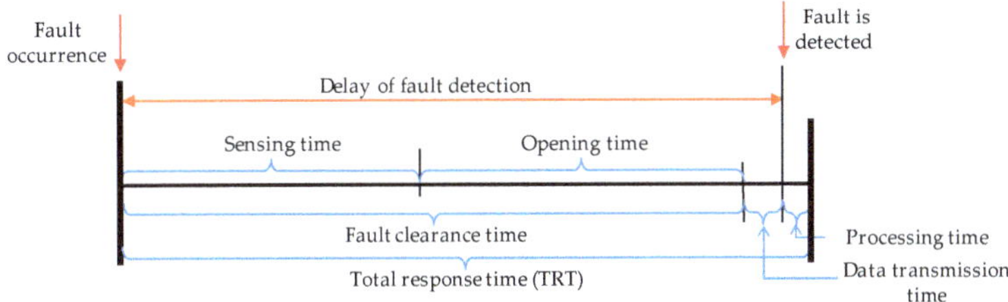

Figure 8. TRT period.

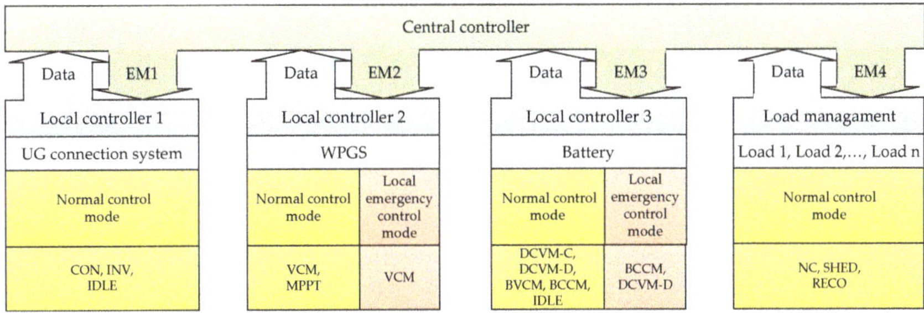

Figure 9. Control structure including LECM for DCV restoration.

Figure 10a shows the proposed DCV restoration algorithm by the LC2 in the WPGS. In this figure, whereas EM2 denotes the EM from the CC, EM2* denotes the final operating mode which will be applied for converter 2. During the normal control mode, LC2 is operated in the MPPT or VCM corresponding to EM from the CC. Even if the grid fault occurs, the CC fails to assign appropriate operating mode since it cannot recognize the grid fault due to the delay of grid fault detection, which results in a rapid increase of the DCV. As soon as the DCV reaches the predefined maximum level of V_{DC}^{max}, the LECM is activated to impose VCM to EM2* regardless of EM from the CC. As a result, the operation of the WPGS is immediately switched into VCM to restore the nominal DCV quickly. As shown in Figure 10a, when the LECM is activated, flag *F* is set to 1 to indicate the LECM operation. Before the CC recognizes the grid fault due to the delay of grid fault detection, the CC does not change EM2. In this case, flag *F* is still kept to 1, and consequently, the LECM assigns EM2* to VCM, which indicates that the DCV is regulated with VCM by converter 2 during period of detection delay. Only when the CC finally recognizes the grid fault, the CC assigns a new EM2 to LC2 according to the PFCS in Figure 2. By detecting the change of EM2 from the CC, LC2 resets flag *F* to 0 to terminate the LECM. Then LC2 starts the normal control mode again by assigning EM2* to new EM2 from the CC.

Figure 10b shows the proposed DCV restoration algorithm by the LC3 in the battery. In this figure, whereas EM3 denotes the EM from the CC, EM3* denotes the final operating mode which will be applied for converter 3. During the normal control mode, LC3 is operated with one among five operating modes according to the EM from the CC. Similarly, in case of the delay of grid fault detection, the DCV can be rapidly decreased due to the power imbalance. As soon as the DCV becomes lower than the first predefined minimum level of V_{DC}^{min1}, the LECM is triggered to restore the DCV. The restoration of the DCV based on the LECM by LC3 is composed of two stages. During the first stage, flag *F1* is set to 1 and EM3* is assigned to BCCM, in which the battery is discharged with the maximum discharging current. By using this discharging mode, the DCV can be recovered toward the nominal value as quickly as possible. As the DCV approaches the second predefined minimum level of V_{DC}^{min2} which is set to higher level than V_{DC}^{min1}, the second stage of the LECM is started. During this stage, F2 is set to 1 and EM3* is assigned to DCVM-D, in which the battery regulates the DCV to restore it completely to the nominal value. Similarly, before the CC recognizes the grid fault due to the delay of grid fault detection, flags F1 and F2 are still kept to 1, and consequently, the LECM assigns EM3* to DCVM-D, which indicates that the DCV is regulated with DCVM-D by converter 3 during period of detection delay. Only when the CC finally recognizes the grid fault, the CC assigns a new EM3 to LC3 according to the PFCS in Figure 2. By detecting the change of EM3 from the CC, LC3 resets flags F1 and F2 to 0 to return back to the normal control mode from the LECM. Then, LC3 starts the normal control mode again by assigning EM3* to new EM3 from the CC.

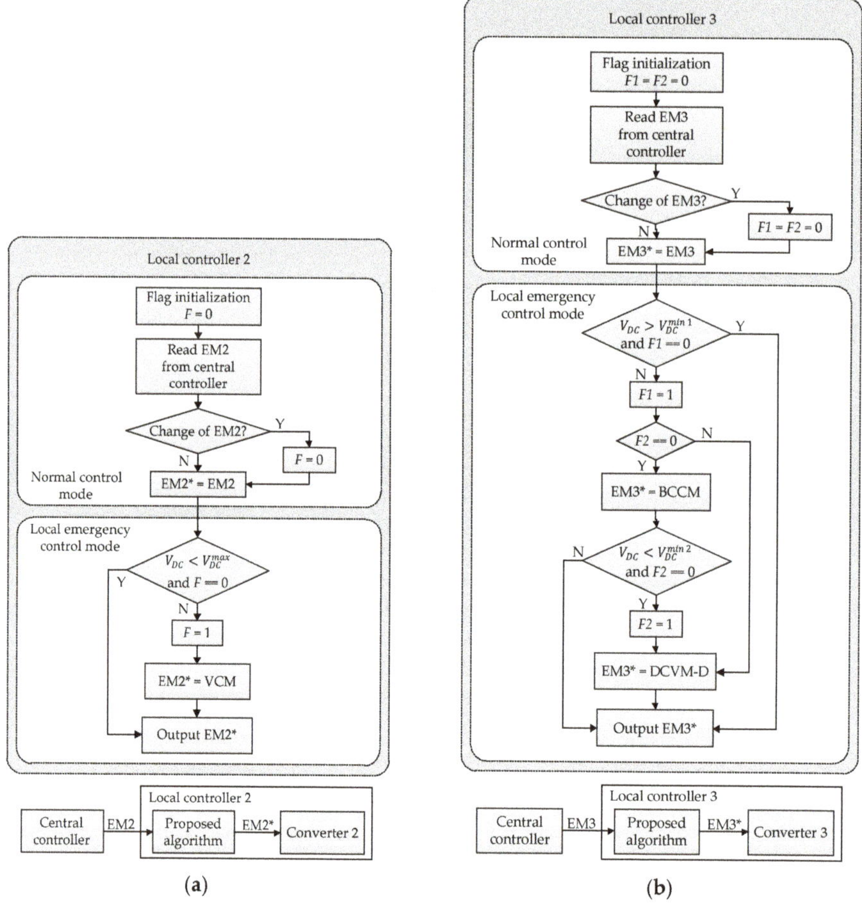

Figure 10. Proposed DCV restoration algorithms. (**a**) By LC2; (**b**) By LC3.

5. Simulation Results

In order to validate the effectiveness of the power flow strategy and the proposed DCV restoration algorithm, simulations were carried out using the PSIM software. Table 3 lists the system parameters used for simulations. Simulations are carried out in three cases which are the grid-connected case, the islanded case, and the case of grid fault detection delay.

Table 3. System parameters of DCMG.

Unit	Parameter	Symbol	Value
UG	Utility grid voltage	V_G^{rms}	220 V
	Grid frequency	F_G	60 Hz
	Inverter-side inductance of LCL filter	L_2	1.7 mH
	Grid-side inductance of LCL filter	L_1	0.9 mH
	Filter capacitance of LCL filter	C_f	4.5 µF
	Transformer Y/Δ	T	380/220 V
PMSG for wind turbine	Stator resistance	R_S	0.64 Ω
	dq-axis inductance	L_{dq}	0.82 mH
	Number of pole pairs	P	6
	Inertia	J	0.111 kgm^2
	Flux linkage	ψ	0.18 Wb
Battery	Maximum discharging power	$P_{B,dis}^{max}$	2 kW
	Maximum voltage	V_B^{max}	265 V
	Maximum charging current	$I_{B,cha}^{max}$	6 A
	Maximum discharging current	$I_{B,dis}^{max}$	8 A
	Maximum *SOC*	SOC^{max}	95 %
	Minimum *SOC*	SOC^{min}	20 %
	Rated capacity	C	30 Ah
DC-link	Voltage	V_{DC}	400 V
	Capacitance	C_{DC}	4 mF
	Maximum DCV	V_{DC}^{max}	420 V
	First minimum DCV	V_{DC}^{min1}	370 V
	Second minimum DCV	V_{DC}^{min2}	390 V
Load	Power of load 1	P_{L1}	0.7 kW
	Power of load 2	P_{L2}	0.5 kW
	Power of load 3	P_{L3}	0.3 kW
	Power of load 4	P_{L4}	0.4 kW
	Power of load 5	P_{L5}	0.8 kW
	Shedding time delay for load 1	T_{she1}	15 ms
	Shedding time delay for load 2	T_{she2}	30 ms
	Shedding time delay for load 3	T_{she3}	45 ms
	Shedding time delay for load 4	T_{she4}	60 ms
	Shedding time delay for load 5	T_{she5}	75 ms
	Time delay for LR	T_{rec}	15 ms
	Priority level: Load 1 < Load 2 < Load 3 < Load 4 < Load 5		

5.1. Grid-Connected Case

In the grid-connected case, the UG is used as the main source to maintain the system power balance under different conditions of wind power, battery status, and load demand. The operation conditions used for the simulation are listed in Table 4.

Table 4. Operation conditions for simulation test in grid-connected case.

	Operation Conditions	Time (s)
1	Load 4 is switched in.	1
2	Wind power increases from 0.25 kW to 2.2 kW.	1.5
3	Load 3 is switched out.	2
4	Wind power increases from 2.2 kW to 3.5 kW.	2.5
5	Battery voltage reaches its maximum value.	3

Figure 11 shows the simulation results for the grid-connected case according to the operation conditions in Table 4. Initially, the system runs stably in operating mode 1 at t = 0.5 s. At this instant, the WPGS works in the MPPT mode to deliver approximately 0.25 kW of wind power to DCMG, the battery is charged with a current of 3 A, and two loads (load 2 and load 3) are connected consuming the total power of 0.8 kW. The UG supplies the power deficit to maintain the system power balance.

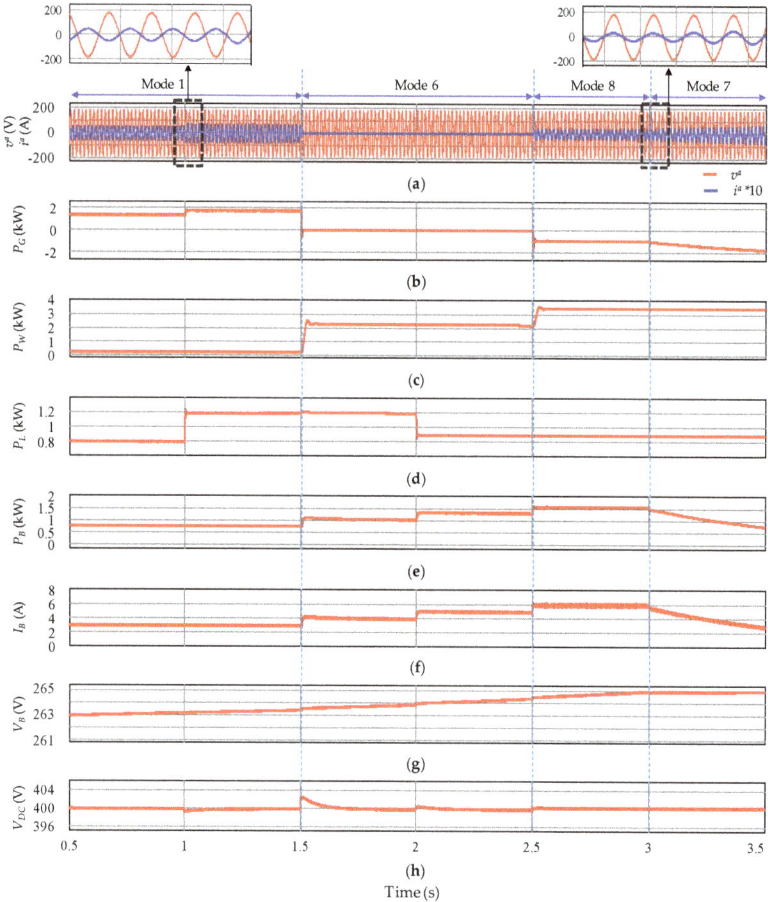

Figure 11. Simulation results for grid-connected case. (**a**) *a*-phase grid voltage and current; (**b**) Output power of UG connection system; (**c**) Output power of WPGS; (**d**) Load power; (**e**) Battery power; (**f**) Battery current; (**g**) Battery voltage; (**h**) DCV.

At t = 1 s, the load demand is suddenly increased because load 4 is switched in. As shown in Figure 11b, the UG increases the supply power to DC-link via REC mode operation of converter 1 to balance the system power exchange. At t = 1.5 s, the generated power from the WPGS is suddenly changed from 0.25 kW to 2.2 kW, which is higher than the load demand (P_L = 1.2 kW). According to the PFCS in Figure 2 and Table 1, the system operation is switched into operating mode 6, in which the battery starts DCVM-C mode to absorb the surplus power in DCMG. In addition, the battery is in charge of regulating the DCV in this operating mode, while the UG is released to IDLE mode. At t = 2 s, load 3 is switched out, which decreases the total load demand to 0.9 kW. Even in this case, the battery keeps balancing the power exchange in DCMG by increasing the charging current as seen in Figure 11f.

When the generated wind power is further increased to 3.5 kW at t = 2.5 s and this increased power is not acceptable to battery since the charging current cannot be increased due to the limitation of $I^{max}_{B,cha}$, the UG connection system is switched from IDLE to INV mode to maintain the system power balance. In this condition, the battery is charged with the maximum charging current of 6 A, in order to avoid overheating and damage. As the battery maximum voltage is reached at around 3 s, the battery operation is changed to BVCM and the power absorbed by battery is gradually reduced. In contrast, the power injected into the UG is gradually increased to keep the system power balance as shown in Figure 11b. These simulation results clearly demonstrate the coordination of the UG and battery during the grid-connected mode.

5.2. Islanded Case

In the islanded case, the DCV is regulated by the coordination operation of the battery and WPGS. Operation conditions used for this simulation are listed in Table 5.

Table 5. Operation conditions for simulation test in islanded case.

	Operation Conditions	Time (s)
1	Grid fault occurs.	1
2	Load 5 is switched in.	1.5
3	Wind power decreases from 1.1 kW to 0.2 kW.	2
4	Wind power increases from 0.2 kW to 1.5 kW.	2.5
5	Wind power increases from 1.5 kW to 3.1 kW.	3
6	Wind power increases from 3.1 kW to 5.0 kW.	3.5

Figure 12 shows the simulation results for the islanded case according to the operation conditions in Table 5. Initially, the system is assumed to operate stably in operating mode 3, in which the DCV is regulated by the UG via REC mode of converter 1. The WPGS is in the MPPT mode, providing approximately 1.1 kW of power to DCMG. Also, it is assumed that the battery is in IDLE mode with the maximum *SOC*, and DC loads consist of load 1, load 2, load 3, and load 4, which consume the total power of 1.9 kW. w t = 1 s, the grid fault happens and DCMG operation is changed into the islanded mode. If the grid fault is quickly detected, the system operation is changed to operating mode 4, and the battery starts discharging to control the DCV with DCVM-D mode. At t = 1.5 s, load 5 is switched in and the battery increases the discharging power to supply extra load demand as seen in Figure 12e. When the wind power is reduced to 0.2 kW at t = 2 s, the battery tries to increase the discharging current to compensate for the power deficit. In this condition, however, the battery cannot control the DCV any longer due to the discharging current limit of 8 A. In order to prevent DCMG system from collapsing even in this case, operating mode 5 is activated with the LS algorithm. As shown in Figure 6, as soon as the counter C_{she} reaches the shedding time delay for load 1, T_{she1} defined in Table 3, load 1 which has the lowest priority is disconnected. Operating mode 5 lasts only during a small duration and the behavior of battery current during this interval is also shown in the magnified figure in Figure 12. After disconnecting load 1, the total load demand remains at 2 kW. On the other hand, the possible supply power consisting of the WPGS power and the maximum discharging battery power is

$$P_W + P^{max}_{B,dis} = 0.2 + 2 = 2.2 \text{ kW} \tag{2}$$

which is higher than the remaining load demand. Therefore, the system operation can be returned to operating mode 4, in which the battery stably controls the DCV after the LS. As the wind power increases from 0.2 kW to 1.5 kW at t = 2.5 s, the available power on DC-link is calculated from (1) as

$$P^{avail}_{DC} = P_W + P^{max}_{B,dis} - P_L = 1.5 + 2 - 2 = 1.5 \text{ kW}. \tag{3}$$

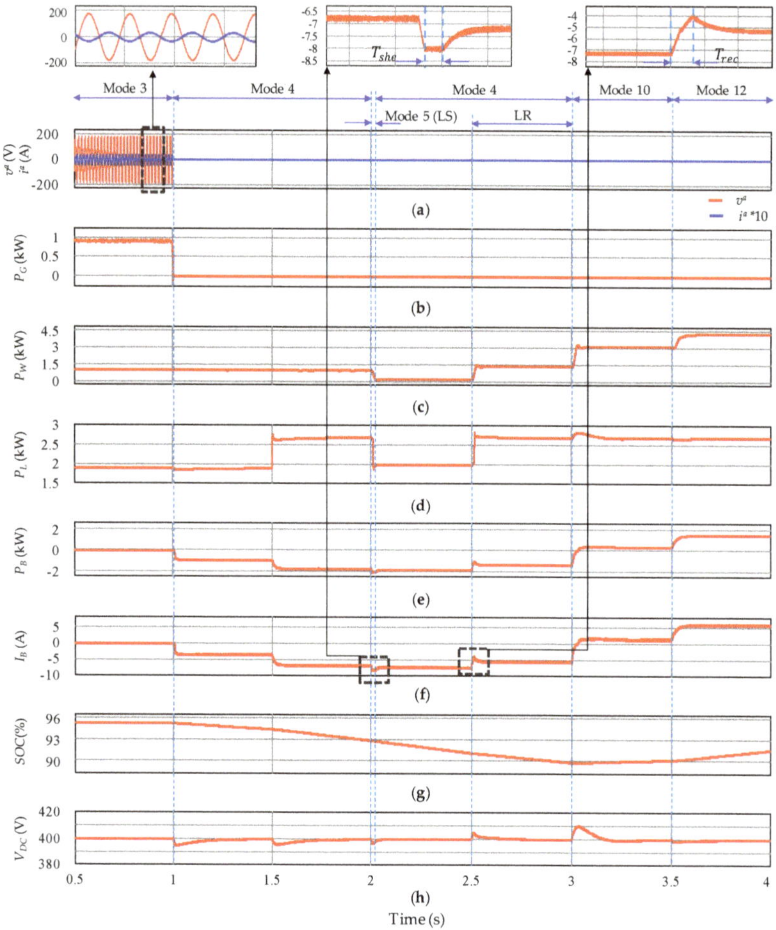

Figure 12. Simulation results for islanded case. (**a**) *a*-phase grid voltage and current; (**b**) Output power of UG connection system; (**c**) Output power of WPGS; (**d**) Load power; (**e**) Battery power; (**f**) Battery current; (**g**) Battery *SOC*; (**h**) DCV.

Because P_{DC}^{avail} is larger than the power of load 1 (0.7 kW), the LR can be activated to reconnect load 1 by using the LR algorithm in Figure 7 and the time delay of T_{rec} defined in Table 3.

At t = 3 s, as the extracted power from the WPGS is increased to 3.1 kW, which is greater than the power demand of load, the system operation is switched into operating mode 10. Accordingly, the battery operating mode is switched from DCVM-D to DCVM-C to absorb the surplus power. At t = 3.5 s, the WPGS injects the power of 5 kW to DCMG. To maintain the power balance of DCMG system, the battery should continue to absorb the excess power by means of DCVM-C operating mode. However, too large excess power results in the battery overcharging current beyond the maximum value of 6 A. To protect the battery from overheating and damage, the system operation is changed to operating mode 12, in which the battery is charged with the maximum limit of 6 A. Because the battery absorbs only a portion of surplus power in this condition, the WPGS is switched from MPPT to VCM to reduce the power extracted from wind, and consequently, to guarantee the system power balance. As shown in Figure 12, the DCV is effectively regulated and the PFCS satisfactorily works in the presence of variations in wind power and load demand, even during islanded mode.

In order to further demonstrate the effectiveness of the PFCS, a comparison in terms of the DCV regulation performance between this paper and the study of [8] is shown in Table 6. In Table 6, the maximum voltage deviation ratio of the DCV is calculated as

$$\text{Maximum voltage deviation ratio} = \frac{\text{Maximum voltage deviation}}{\text{Nominal voltage}} \times 100. \quad (4)$$

Table 6. Comparison of the DCV regulation performance.

	Study in Ref [8]	This Paper
Maximum voltage deviation ratio of the DCV in grid-connected case	1.84%	0.75%
Maximum voltage deviation ratio of the DCV in islanded case	5.26%	2.5%

As shown in Table 6, as compared to the simulation results in Ref [8], the DCV can be regulated at the desirable value with significantly smaller voltage deviation ratio in this paper, in spite of the variations in the generated power from the WPGS and load demand.

5.3. Case of Grid Fault Detection Delay

In this section, simulations are carried out to prove the effectiveness of the proposed DCV restoration scheme for the system stability in case of the grid fault detection delay. The DCV restoration can be achieved either by the battery as shown in Figures 13 and 14, or by the WPGS as shown in Figures 15 and 16. In these simulation results, 0.46 s is used for the fault clearance time.

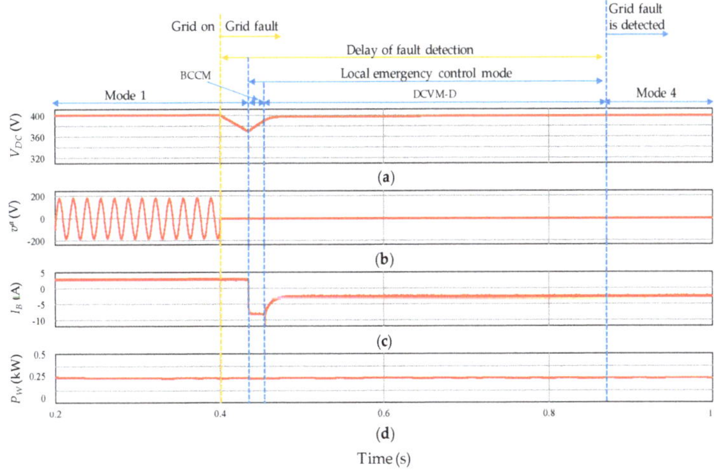

Figure 13. Simulation results of the proposed DCV restoration in case of grid fault detection delay during operating mode 1. (**a**) DCV; (**b**) a-phase grid voltage; (**c**) Battery current; (**d**) Output power of WPGS.

Figures 13 and 14 show the simulation results of the proposed DCV restoration in case of grid fault detection delay during operating mode 1 and operating mode 3, respectively. Before the fault occurs in the UG, DCMG works in operating mode 1 or operating mode 3, in which the system power balance is maintained by the UG via REC mode of converter 1. In both figures, the WPGS is operated in the MPPT mode while the battery is charging in operating mode 1 and is in IDLE mode in operating mode 3. Even though the UG shuts down suddenly at t = 0.4 s, the CC does not recognize it because of the fault detection delay. Consequently, DCMG still operates at operating mode 1 in Figure 13 and operating mode 3 in Figure 14, which results in power imbalance and rapid decrease of the DCV since any power sources do not control the DCV during this period. As soon as the DCV drops to V_{DC}^{min1} of 370 V, LECM by LC3 is activated. At this instant, the battery operation starts BCCM, discharging the maximum current to restore the DCV quickly. When the DCV reaches V_{DC}^{min2} of 390 V, the operating mode of battery is automatically switched into DCVM-D to gradually regulate the DCV at the nominal value of 400 V. Once the grid fault is detected with delay at t = 0.55 s, the CC changes the system operation to operating mode 4, terminating LECM by LC3. When DCMG operation returns to the normal mode, the battery continuously controls the DCV by means of DCVM-D mode, with a seamless transition between LECM and the normal mode.

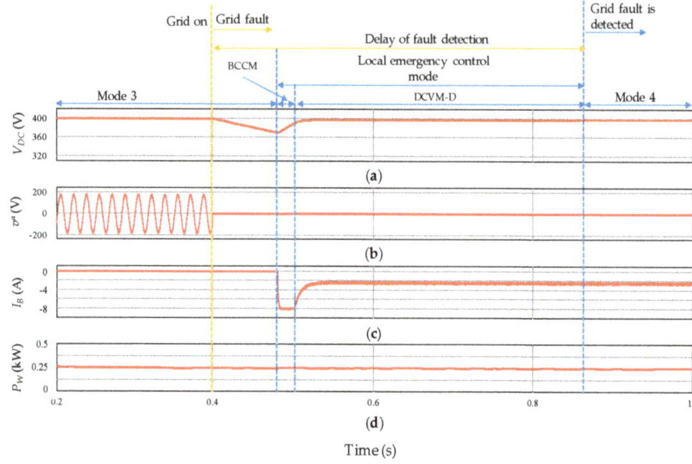

Figure 14. Simulation results of the proposed DCV restoration in case of grid fault detection delay during operating mode 3. (**a**) DCV; (**b**) *a*-phase grid voltage; (**c**) Battery current; (**d**) Output power of WPGS.

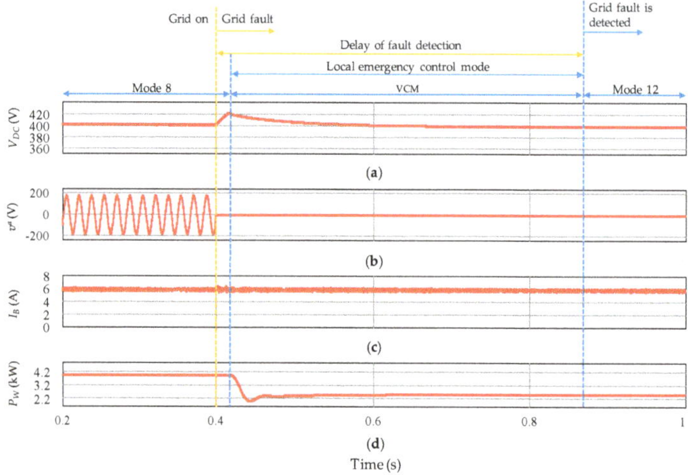

Figure 15. Simulation results of the proposed DCV restoration in case of grid fault detection delay during operating mode 8. (**a**) DCV; (**b**) *a*-phase grid voltage; (**c**) Battery current; (**d**) Output power of WPGS.

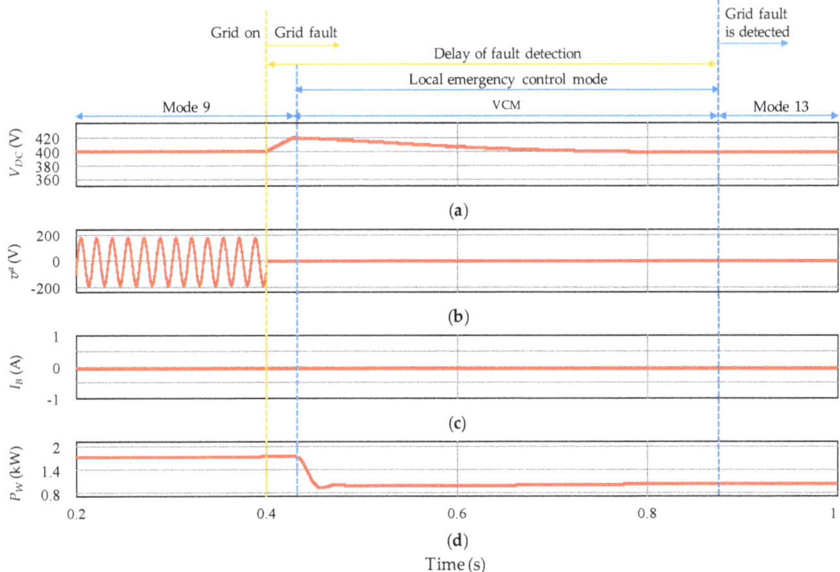

Figure 16. Simulation results of the proposed DCV restoration in case of grid fault detection delay during operating mode 9. (**a**) DCV; (**b**) *a*-phase grid voltage; (**c**) Battery current; (**d**) Output power of WPGS.

Figures 15 and 16 show the simulation results of the proposed DCV restoration in case of grid fault detection delay during operating mode 8 and operating mode 9, respectively. Before DCMG enters the islanded mode due to the fault occurrence, the UG regulates the DCV stably by absorbing surplus power via INV mode of converter 1. In both figures, the WPGS is operated in the MPPT mode while the battery is charged with the maximum charging current in operating mode 8 and is in IDLE mode in operating mode 9. Similarly, the UG has a fault suddenly at t = 0.4 s. In case of the grid fault detection delay, the CC still uses the UG for the DCV regulation. Under this condition, the UG is not able to absorb excess power any longer, a rapid increase of the DCV and power imbalance are introduced in DCMG. When the DCV reaches the maximum level of 420 V, LECM by LC2 is triggered. As a result, the operating mode of the WPGS is instantly changed to VCM to adjust the DCV to the nominal value of 400 V. Once the grid fault is detected with delay, the CC changes the system operation to operating mode 12 in Figure 15 and operating mode 13 in Figure 16, respectively, according to the PFCS shown in Figure 2. This terminates the LECM by LC2, returning to the normal mode in which the WPGS works with VCM with seamless transition between the LECM and the normal mode.

In order to further demonstrate the effectiveness of the proposed DCV restoration algorithm, the simulation results in case of grid fault detection delay during operating mode 3 without the proposed DCV restoration algorithm are shown in Figure 17. As compared with Figure 14, it is shown that the DCV is rapidly dropped due to the delay of grid fault detection without the proposed DCV restoration algorithm in Figure 17, which confirms the effectiveness of the proposed DCV restoration algorithm.

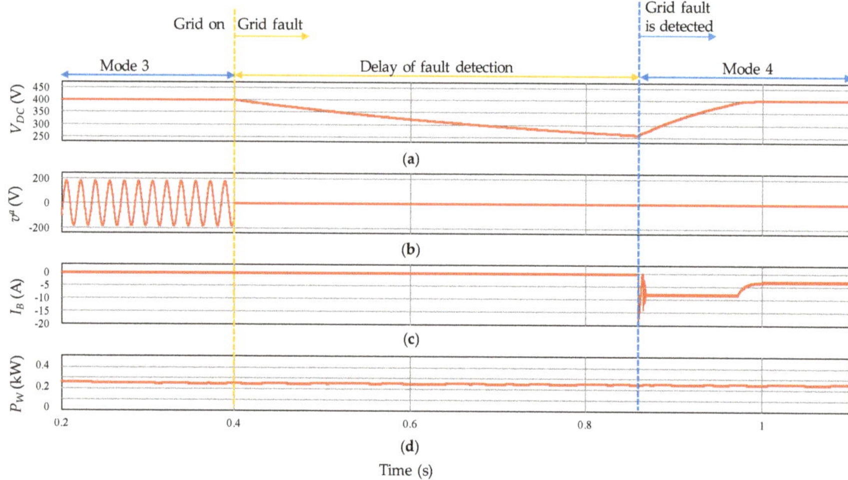

Figure 17. Simulation results in case of grid fault detection delay during operating mode 3 without the proposed DCV restoration algorithm. (**a**) DCV; (**b**) a-phase grid voltage; (**c**) Battery current; (**d**) Output power of WPGS.

6. Experimental Results

In order to validate the effectiveness of the PFCS and the proposed DCV restoration algorithm, experiments based on the laboratory testbed system were carried out. Figure 18 shows the configuration of the experimental system including three power converters to connect the UG, battery, and WPGS in DCMG. This figure also shows a lithium-ion battery and a wind turbine emulator which is constructed by a PMSG and an induction machine. Similar to the simulations, the experiments are performed in three cases: the grid-connected case, islanded case, and the case of grid fault detection delay.

Figure 18. Configuration of the experimental system.

6.1. Grid-Connected Case

Figure 19 shows the experimental results of DCMG operation for the grid-connected case under the variations of load and wind power. Figure 19a presents DCMG behavior under load power variation. Before load variation, DCMG runs stably in operating mode 1, in which the WPGS is operated in the MPPT mode to inject the maximum power from wind to DCMG and the battery is charged with a constant current of 3 A. The DCV regulation and power balance is achieved by the UG through REC mode of converter 1. At this instant, the load demand is suddenly increased from 0.4 kW to 0.8 kW. To feed extra load demand, the UG increases the supply power to DC-link from 1.2 kW to 1.6 kW as shown in Figure 19a. The battery still works in the same operating mode and the DCV is maintained at the nominal value of 400 V, in spite of load variation.

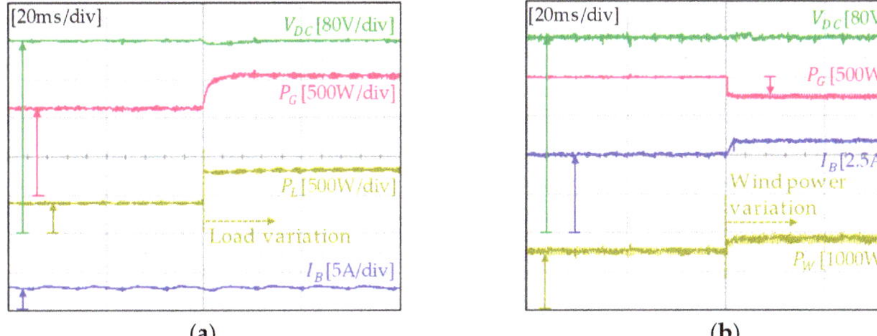

Figure 19. Experimental results for grid-connected case. (**a**) Load variation at mode 1; (**b**) Transition from mode 6 to mode 8.

Figure 19b illustrates the PFCS performance under the mode transition instant. Initially, DCMG is operated in operating mode 6, in which the DCV is controlled by DCVM-C of battery. Meanwhile, the WPGS is running in the MPPT mode and the UG is in IDLE state. When the battery cannot store excess power in DCMG as a result of the increase of wind power from 1.5 kW to 1.9 kW, surplus energy can be used to inject to the UG. In this situation, to ensure the power balance in DCMG, the system enters operating mode 8 and the UG system should be changed from IDLE to INV mode by regulating the DCV. Instead, the battery releases the DCV regulation to the UG system, operating in BCCM with the charging current of 6 A. Despite the wind power variation and resultant operating mode change, it is shown in this figure that the DCV is well maintained at 400 V, with only small transient.

6.2. Islanded Case

Figure 20 shows the experimental results of DCMG operation for three situations in the islanded case. Figure 20a presents transition results from the grid-connected to islanded mode, Figure 20b shows the results of the LS algorithm, and Figure 20c illustrates the results of LR algorithm, respectively. In Figure 20a, DCMG initially operates stably in operating mode 3. The power balance is maintained by the UG and the battery operation is in IDLE mode. When the UG shuts down abnormally, the system operation is switched to operating mode 4, in which the battery starts DCVM-D mode with discharging current of 1.3 A to compensate the power deficit caused by the UG outage. It is confirmed from Figure 20a that the DCV is stably regulated at 400 V, with only small transient, even under transient conditions in islanded mode.

Figure 20b shows the experimental results for the LS algorithm. At first, the system operates stably in operating mode 4, in which the DCV is controlled by DCVM-D mode of battery and the total load demand is 2.25 kW. As the wind power suddenly drops from 0.75 kW to 0.25 kW, the battery increases its discharging current to compensate this power deficit. Due to the maximum discharging current limit, however, the power balance cannot be achieved by battery discharging. In this case, the LS is inevitable to avoid the system collapse and the DCV reduction. As soon as the LS algorithm is started after 15 ms as seen in Figure 20b, the total load demand is reduced to 1.8 kW by disconnecting load of 0.45 kW. As a result, the battery returns to DCVM-D mode again to regulate the DCV continuously.

Figure 20c shows the experimental results for the LR algorithm. This operation may happen when the wind power in DCMG increases again after the LS. If the wind power increases from 0.25 kW to 0.9 kW after the LS, the CC calculates the available power on DC-link as

$$P_{DC}^{avail} = P_W + P_{B,dis}^{max} - P_L = 0.9 + 2 - 1.8 = 1.1 \text{ kW}. \tag{5}$$

Because P_{DC}^{avail} is larger than the disconnected load power (0.45 kW) by the LS, this load can be reconnected again after T_{rec} of 15 ms as shown in Figure 20c. These experimental results coincide well with those of the simulations.

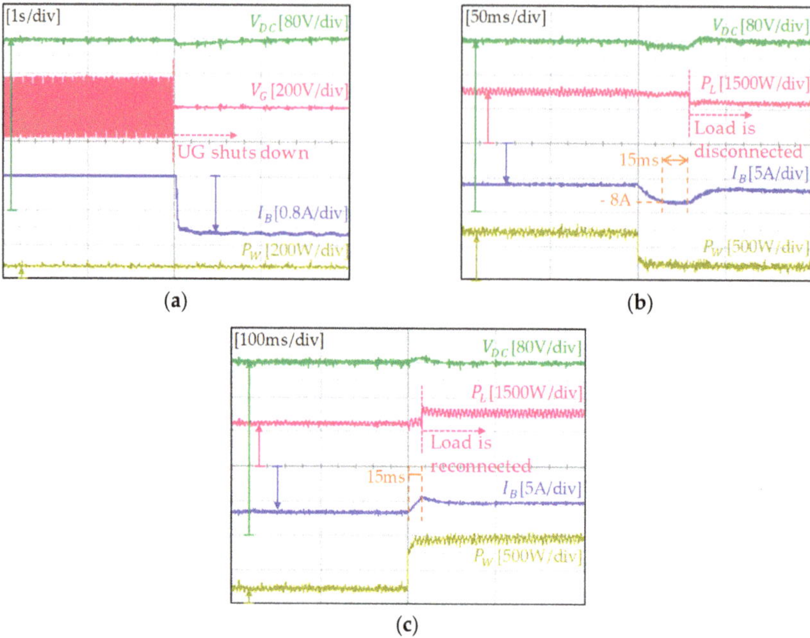

Figure 20. Experimental results for islanded case. (**a**) Transition from grid-connected to islanded mode; (**b**) LS; (**c**) LR.

6.3. Case of Grid Fault Detection Delay

In this section, the experimental results are presented to validate the effectiveness of the proposed DCV restoration scheme. Figure 21 shows the experimental results of the proposed DCV restoration in case of the grid fault detection delay under different operating modes.

In particular, Figure 21a,b show the experimental results of the proposed DCV restoration implemented by using the battery. Before the UG has a fault, the battery is charging with the current of 3 A in operating mode 1 in Figure 21a and is in IDLE in operating mode 3 in Figure 21b, respectively. As soon as the DCV drops to 370 V, LECM is activated and the battery starts BCCM with the discharging current of 8 A. When the DCV is recovered to 390 V, the battery operation is switched into DCVM-D to gradually regulate the DCV at 400 V. As demonstrated, the experimental results in Figure 21a,b are well matched with the simulation results in Figures 13 and 14.

Figure 21c,d show the experimental results of the proposed DCV restoration obtained by using the WPGS. Before the fault occurs in the UG, the battery is charging with the maximum charging current, i.e., 6 A, in operating mode 8 in Figure 21c, and is in IDLE in operating mode 9 in Figure 21d, respectively. Meanwhile, the WPGS works in the MPPT mode to get the maximum power from wind. As explained, the DCV increases rapidly due to the delay of grid fault detection. When the DCV reaches 420 V, LECM is started and the WPGS operation is switched into VCM mode to regulate the DCV at 400 V. These experimental results are also matched well with the simulation results in Figures 15 and 16. As a result, it is confirmed that the DCV which has an essential role in DCMG operation can be restored effectively by using the proposed scheme.

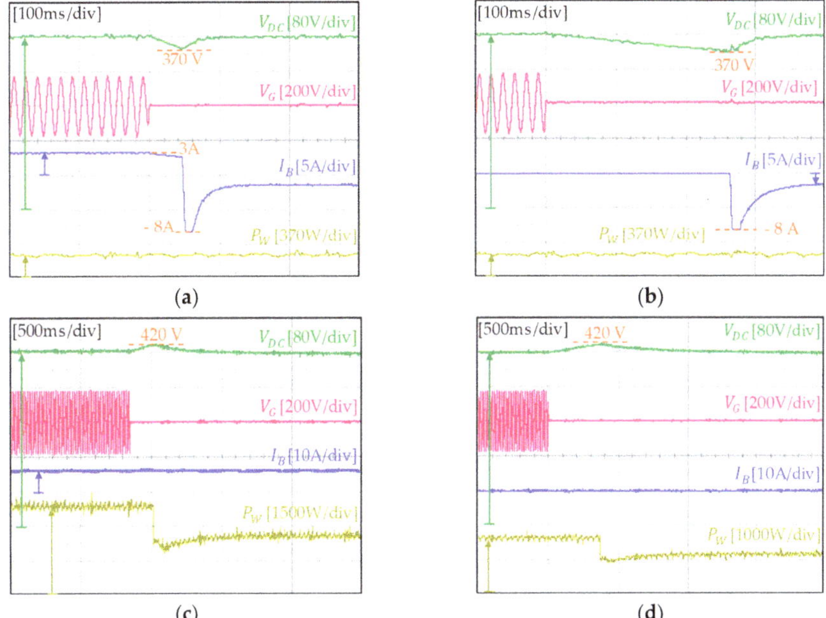

Figure 21. Experimental results of the proposed DCV restoration in case of grid fault detection delay. (**a**) During operating mode 1; (**b**) During operating mode 3; (**c**) During operating mode 8; (**d**) During operating mode 9.

7. Discussion

Based on the simulation and experimental results, the effectiveness of the PFCS and the proposed DCV restoration algorithm for DCMG has been comprehensively validated. The PFCS of DCMG consisting of thirteen operating modes effectively maintains the system power balance both in the grid-connected and islanded modes. Also, the proposed DCV restoration algorithm rapidly recovers the DCV through the LECM operation even in critical cases, which enhances the reliability and expansion of DCMG. However, it is worth noting that the performance of the proposed scheme was evaluated only for DCMG, which is composed of a single battery and a single WPGS. As a result, the proposed scheme can be achieved by a simple control structure. However, since multiple units of ESS and RES are being integrated into DCMGs recently, the implementation of the DCV restoration algorithm in this configuration becomes an important issue in view of the requirement of power-sharing among multiple power systems. DCV restoration in the presence of parallel-connected multiple power systems needs several considerations and modifications in the algorithm; as such, future research works will be devoted to this issue.

8. Conclusions

This paper has presented an effective PFCS based on the centralized control approach and a reliable DCV restoration algorithm for DCMG under grid fault conditions. The main contributions of this paper can be summarized as follows:

(i) By taking into account the relationship of supply-demand power and the statuses of system units such as the UG, WPGS, and battery, the CC generates thirteen proper operating modes to each LC to effectively deal with various conditions in DCMG. As a result, the stability of DCV and the system power balance can be guaranteed. Moreover, by considering both the grid-connected and islanded modes, a comprehensive performance evaluation for the PFCS was undertaken.

(ii) In the PFCS, the LS is realized in order to prevent the system from collapsing in critical cases, and the LR is also implemented to reconnect the loads which are disconnected due to power shortages in DCMG. In both the LS and LR algorithms, the time delay is used to avoid undesirable load disconnections or reconnections caused by noises.
(iii) To prevent the system power imbalance in DCMG caused by the delay of grid fault detection, a reliable DCV restoration algorithm based on the LECM by the LCs of the WPGS and battery-based ESS is also proposed in this paper. By using the proposed scheme, the WPGS or battery instantly starts LECM to restore the DCV rapidly to the nominal value, regardless of the control signals from the CC, as soon as abnormal behavior of the DCV is detected. The experimental tests demonstrate that the proposed scheme restores the DCV to the nominal value within 0.5 s, even in the worst case. Furthermore, after the UG fault is recognized, the proposed scheme transfers DCMG operation from LECM to the normal operating mode, seamlessly and stably.

In order to evaluate the feasibility of the PFCS and the proposed DCV restoration algorithm, a prototype DCMG, consisting of three-phase grid-connected converter, WPGS, and lithium-ion battery-based ESS, was constructed by using 32-bit floating-point digital signal processor (DSP) TMS320F28335 controller. In the experimental setup, the wind turbine emulator consisting of a coupled PMSM-inductor machine and three-phase converter was employed for the WPGS. Comprehensive simulation and experimental results based on the PSIM and DCMG testbed are presented to verify the effectiveness of the PFCS and the proposed DCV restoration algorithm.

Author Contributions: T.V.N. and K.-H.K. conceived the main concept of the control structure and developed the entire system. T.V.N. carried out the research and analyzed the numerical data with guidance from K.-H.K. T.V.N. and K.-H.K. collaborated in the preparation of the manuscript.

Funding: "This research was funded by Seoul Tech (Seoul National University of Science and Technology)" and "The APC was funded by Seoul Tech (Seoul National University of Science and Technology)".

Acknowledgments: This study was supported by the Advanced Research Project funded by the SeoulTech (Seoul National University of Science and Technology).

Conflicts of Interest: The authors declare no conflict of interest.

Abbreviations

AC	Alternating Current
ACMG	AC Microgrid
BCCM	Battery Current Control Mode
BVCM	Battery Voltage Control Mode
CC	Central Controller
CC-CV	Current Control-Voltage Control
DBS	DC-Bus Signaling
DC	Direct Current
DCMG	DC Microgrid
DCV	DC-link Voltage
DCVM-C	DC-link Voltage control Mode by battery Charging
DCVM-D	DC-link Voltage control Mode by battery Discharging
DG	Distributed Generation
DIS	Disable
DSP	Digital Signal Processor
EM	Execution Mode
ESS	Energy Storage System
INV	Inverter
LC	Local Controller
LECM	Local Emergency Control Mode
LR	Load Reconnection

LS	Load Shedding
MG	Microgrid
LR	Load Reconnection
LS	Load Shedding
MG	Microgrid
MPPT	Maximum Power Point Tracking
NC/RECO	No change/Reconnection
PFCS	Power Flow Control Strategy
PI	Proportional Integral
PMSG	Permanent Magnet Synchronous Generator
PWM	Pulse Width Modulation
REC	Rectifier
RES	Renewable Energy Source
SHED	Shedding
SOC	State of Charge
SVM	Space Vector Modulation
TRT	Total Response Time
UG	Utility Grid
VCM	Voltage Control Mode
WPGS	Wind Power Generation System
C	Rated capacity of battery
C_{rec}	Counter of load reconnection
C_{she}	Counter of load shedding
d	Duty cycle
EM2	Execution mode from the CC to converter 2 (WPGS)
EM2*	Final operating mode applied for converter 2 (WPGS)
EM3	Execution mode from the CC to converter 3 (battery)
EM3*	Final operating mode applied for converter 3 (battery)
F	Flag in the LECM of WPGS
$F1, F2$	Flag in the LECM of battery
i	Load status indication
i^a	a-phase grid current
I_B	Battery current
I_B^{ref}	Battery current reference
$I_{B,cha}^{max}$	Maximum battery charging current
$I_{B,cha}^{req}$	Required battery charging current
$I_{B,dis}^{max}$	Maximum battery discharging current
$I_{B,dis}^{req}$	Required battery discharging current
I_d	d-axis current
I_d^{ref}	d-axis current reference
I_q	q-axis current
I_q^{ref}	q-axis current reference
LC2	Local controller in the WPGS
LC3	Local controller in the battery
n	Total quantity of load
P_B	Battery power
$P_{B,dis}^{max}$	Maximum discharging power of battery
P_{DC}^{avail}	Available power on DC-link
P_G	Output power of UG connection system
P_L	Load power
P_W	Output power of WPGS
SOC^{max}	Maximum battery state of charge
SOC^{min}	Minimum battery state of charge
T_{rec}	Time delay for load reconnection

T_{she}	Time delay for load shedding
v^a	*a*-phase grid voltage
V_B	Battery voltage
V_B^{max}	Maximum battery voltage
V_{DC}	DC-link voltage
V_{DC}^{max}	Maximum DC-link voltage in LECM of WPGS
V_{DC}^{min1}	First minimum DC-link voltage in LECM of battery
V_{DC}^{min2}	Second minimum DC-link voltage in LECM of battery
V_{DC}^{ref}	DC-link voltage reference
ω	Generator rotor speed
ω^{ref}	Generator rotor speed reference
ΔT	Step size

References

1. Bose, B.K. Global Warming: Energy, Environmental Pollution, and the Impact of Power Electronics. *IEEE Ind. Electron. Mag.* **2010**, *4*, 6–17. [CrossRef]
2. Olivares, D.E.; Mehrizi-Sani, A.; Etemadi, A.H.; Cañizares, C.A.; Iravani, R.; Kazerani, M.; Hajimiragha, A.H.; Gomis-Bellmunt, O.; Saeedifard, M.; Palma-Behnke, R.; et al. Trends in Microgrid Control. *IEEE Trans. Smart Grid* **2014**, *5*, 1905–1919. [CrossRef]
3. Molderink, A.; Bakker, V.; Bosman, M.G.C.; Hurink, J.L.; Smit, G.J.M. Management and Control of Domestic Smart Grid Technology. *IEEE Trans. Smart Grid* **2010**, *1*, 109–119. [CrossRef]
4. Ma, W.; Wang, W.; Wu, X.; Hu, R.; Tang, F.; Zhang, W. Control Strategy of a Hybrid Energy Storage System to Smooth Photovoltaic Power Fluctuations Considering Photovoltaic Output Power Curtailment. *Sustainability* **2019**, *11*, 1324. [CrossRef]
5. Salas-Puente, R.A.; Marzal, S.; González-Medina, R.; Figueres, E.; Garcera, G. Power Management of the DC Bus Connected Converters in a Hybrid AC/DC Microgrid Tied to the Main Grid. *Energies* **2018**, *11*, 794. [CrossRef]
6. Han, Y.; Chen, W.; Li, Q. Energy Management Strategy Based on Multiple Operating States for a Photovoltaic/Fuel Cell/Energy Storage DC Microgrid. *Energies* **2017**, *10*, 136. [CrossRef]
7. Dragičević, T.; Guerrero, J.M.; Vasquez, J.C.; Škrlec, D. Supervisory Control of an Adaptive-Droop Regulated DC Microgrid with Battery Management Capability. *IEEE Trans. Power Electron.* **2014**, *29*, 695–706. [CrossRef]
8. Gao, L.; Liu, Y.; Ren, H.; Guerrero, J.M. A DC Microgrid Coordinated Control Strategy Based on Integrator Current-Sharing. *Energies* **2017**, *10*, 1116. [CrossRef]
9. Jeong, D.-K.; Kim, H.-S.; Baek, J.-W.; Kim, H.-J.; Jung, J.-H. Autonomous Control Strategy of DC Microgrid for Islanding Mode Using Power Line Communication. *Energies* **2018**, *11*, 924. [CrossRef]
10. Mohammadi, F. Power Management Strategy in Multi-Terminal VSC-HVDC System. In Proceedings of the 4th National Conference on Applied Research in Electrical, Mechanical, Computer and IT Engineering, Tehran, Iran, 4 October 2018.
11. Lee, S.-J.; Choi, J.-Y.; Lee, H.-J.; Won, D.-J. Distributed Coordination Control Strategy for a Multi-Microgrid Based on a Consensus Algorithm. *Energies* **2017**, *10*, 1017. [CrossRef]
12. Nguyen, T.-L.; Guillo-Sansano, E.; Syed, M.H.; Nguyen, V.-H.; Blair, S.M.; Reguera, L.; Tran, Q.-T.; Caire, R.; Burt, G.M.; Gavriluta, C.; et al. Multi-Agent System with Plug and Play Feature for Distributed Secondary Control in Microgrid—Controller and Power Hardware-in-the-Loop Implementation. *Energies* **2018**, *11*, 3253. [CrossRef]
13. Yue, J.; Hu, Z.; Li, C.; Vasquez, J.C.; Guerrero, J.M. Economic Power Schedule and Transactive Energy through an Intelligent Centralized Energy Management System for a DC Residential Distribution System. *Energies* **2017**, *10*, 916. [CrossRef]
14. Kaur, A.; Kaushal, J.; Basak, P. A Review on Microgrid Central Controller. *Renew. Sustain. Energy Rev.* **2016**, *55*, 338–345. [CrossRef]
15. Dragičević, T.; Lu, X.; Vasquez, J.C.; Guerrero, J.M. DC Microgrids-Part I: A Review of Control Strategies and Stabilization Techniques. *IEEE Trans. Power Electron.* **2016**, *31*, 4876–4891.

16. Geng, Y.; Hou, M.; Zhang, L.; Dong, F.; Jin, Z. An Improved Voltage Control Strategy for DC Microgrid with Hybrid Storage System. In Proceedings of the 2018 13th IEEE Conference on Industrial Electronics and Applications (ICIEA), Wuhan, China, 31 May–2 June 2018; pp. 958–962.
17. Sanjeev, P.; Padhy, N.P.; Agarwal, P. Peak Energy Management Using Renewable Integrated DC Microgrid. *IEEE Trans. Smart Grid* **2018**, *9*, 4906–4917. [CrossRef]
18. Salas-Puente, R.; Marzal, S.; González-Medina, R.; Figueres, E.; Garcera, G. Experimental Study of a Centralized Control Strategy of a DC Microgrid Working in Grid Connected Mode. *Energies* **2017**, *10*, 1627. [CrossRef]
19. Liu, B.; Zhuo, F.; Zhu, Y.; Yi, H. System Operation and Energy Management of a Renewable Energy-Based DC Micro-Grid for High Penetration Depth Application. *IEEE Trans. Smart Grid* **2015**, *6*, 1147–1155. [CrossRef]
20. Chen, M.; Ma, S.; Wan, H.; Wu, J.; Jiang, Y. Distributed Control Strategy for DC Microgrids of Photovoltaic Energy Storage Systems in Off-Grid Operation. *Energies* **2018**, *11*, 2637. [CrossRef]
21. Sanjeev, P.; Padhy, N.P.; Agarwal, P. A New Architecture for DC Microgrids using Supercapacitor. In Proceedings of the 2018 9th IEEE International Symposium on Power Electronics for Distributed Generation Systems (PEDG), Charlotte, NC, USA, 25–28 June 2018; pp. 1–5.
22. Li, F.; Li, R.; Zhou, F. *Microgrid Technology and Engineering Application*, 1st ed.; Elsevier: London, UK, 2015; pp. 59–60.
23. Sanjeev, P.; Padhy, N.P.; Agarwal, P. Autonomous Power Control and Management between Standalone DC Microgrids. *IEEE Trans. Ind. Inform.* **2018**, *14*, 2941–2950. [CrossRef]
24. Xu, L.; Chen, D. Control and Operation of a DC Microgrid with Variable Generation and Energy Storage. *IEEE Trans. Power Deliv.* **2011**, *26*, 2513–2522. [CrossRef]
25. Khamis, A.; Xu, Y.; Dong, Z.Y.; Zhang, R. Faster Detection of Microgrid Islanding Events using an Adaptive Ensemble Classifier. *IEEE Trans. Smart Grid* **2018**, *9*, 1889–1899. [CrossRef]
26. Kim, M.-S.; Haider, R.; Cho, G.-J.; Kim, C.-H.; Won, C.-Y.; Chai, J.-S. Comprehensive Review of Islanding Detection Methods for Distributed Generation Systems. *Energies* **2019**, *12*, 837. [CrossRef]
27. Ahmad, K.N.E.K.; Selvaraj, J.; Rahim, N.A. A Review of the Islanding Detection Methods in Grid-Connected PV Inverters. *Renew. Sustain. Energy Rev.* **2013**, *21*, 756–766. [CrossRef]
28. Zhang, J.; Lai, J.-S.; Kim, R.-Y.; Yu, W. High-Power Density Design of a Soft-Switching High-Power Bidirectional dc–dc Converter. *IEEE Trans. Power Electron.* **2007**, *22*, 1145–1153. [CrossRef]
29. Wong, Y.S.; Hurley, W.G.; Wölfle, W.H. Charge Regimes for Valve-Regulated Lead-Acid Batteries: Performance Overview Inclusive of Temperature Compensation. *J. Power Sources* **2008**, *183*, 783–791. [CrossRef]
30. Bahrani, B.; Karimi, A.; Rey, B.; Rufer, A. Decoupled dq-Current Control of Grid-Tied Voltage Source Converters Using Nonparametric Models. *IEEE Trans. Ind. Electron.* **2013**, *60*, 1356–1366. [CrossRef]
31. IEEE. *IEEE Standard for Interconnection and Interoperability of Distributed Energy Resources with Associated Electric Power Systems Interfaces*; IEEE Std.1547; IEEE: New York, NY, USA, 2018.

Article

A Bidirectional Power Charging Control Strategy for Plug-in Hybrid Electric Vehicles

Fazel Mohammadi [1,*], Gholam-Abbas Nazri [2] and Mehrdad Saif [1]

1 Electrical and Computer Engineering (ECE) Department, University of Windsor, Windsor, ON N9B 1K3, Canada
2 Electrical and Computer Engineering, Wayne State University, Detroit, MI 48202, USA
* Correspondence: fazel@uwindsor.ca or fazel.mohammadi@ieee.org

Received: 17 May 2019; Accepted: 26 July 2019; Published: 9 August 2019

Abstract: Plug-in Hybrid Electric Vehicles (PHEVs) have the potential of providing frequency regulation due to the adjustment of power charging. Based on the stochastic nature of the daily mileage and the arrival and departure time of Electric Vehicles (EVs), a precise bidirectional charging control strategy of plug-in hybrid electric vehicles by considering the State of Charge (SoC) of the batteries and simultaneous voltage and frequency regulation is presented in this paper. The proposed strategy can control the batteries charge which are connected to the grid, and simultaneously regulate the voltage and frequency of the power grid during the charging time based on the available power when different events occur over a 24-h period. The simulation results prove the validity of the proposed control strategy in coordinating plug-in hybrid electric vehicles aggregations and its significant contribution to the peak reduction, as well as power quality improvement. The case study in this paper consists of detailed models of Distributed Energy Resources (DERs), diesel generator and wind farm, a generic aggregation of EVs with various charging profiles, and different loads. The test system is simulated and analyzed in MATLAB/SIMULINK software.

Keywords: bidirectional power flow; charging station; Distributed Generations (DGs); microgrids; Plug-in Electric Vehicles (PHEVs); State of Charge (SoC)

---

1. Introduction

1.1. Motivation

The rapid increase in energy demand, destruction of the earth's resources, and discharge of carbon dioxide are the leading causes of environmental pollution and climate change in the world. Further, transportation is more attentive, considering the fact that it causes more than 15% of carbon dioxide discharge, which is critical for all the people [1,2]. As a result of that, the transition from the Internal Combustion Engine (ICE) to hybrid and full-electric vehicles has been an immense focus for the reduction of greenhouse gases [2]. Due to pollution and energy crisis, many studies in the field of Electric-Drive Vehicles (EDVs), including Battery Electric Vehicles (BEVs), Hybrid Electric Vehicles (HEVs), and Plug-in Hybrid Electric Vehicles (PHEVs) have appeared worldwide [2–4]. However, developments in the field of electric vehicles are restricted by the technology of the batteries and Energy Storage Systems (ESSs), there are many positive signs of progress in this field [5–7]. A good number of alternative energy sources, which are renewable, are already being harnessed and utilized to meet the energy demand in the world [8,9]. This paper mainly focuses on PHEVs to study the impact of EVs and their interconnection to the power system.

The arbitrary connection of PHEVs to the power grid leads to the complicated operation, planning, and control of the power system. There are different charging mechanisms for PHEVs to be charged. IEC Std. 61851 is one of the common standards, which is established for PHEVs charging [8]. Regardless of the charging mechanism, the availability of charging stations is an essential factor that should be considered for power system control, operation, and long-term planning. The common and low-cost procedure for charging PHEVs is slow charging, which drains less power from the grid. However, the main drawback of this mechanism is that it takes more time to fully charge the batteries of PHEVs. On the other hand, the fast charging mechanism, based on the new developments in power electronics devices and restructuring both of both ESSs and chargers, can expedite the charging process and charge a depleted battery from 10 to 80 percent in half an hour [10]. However, this mechanism drains more current at a high voltage level from the network and has a negative effect on the other loads connected to the same bus, e.g., the voltage drop at the end of the power line.

In modern power systems, microgrids are defined as interconnected local energy centers with control and management capabilities and clear boundaries. They enable bidirectional and autonomous power exchange to prevent power outage by providing high-quality operation and more reliable energy supply to the load centers. Aging electric power grid infrastructures, continuous increase in the load demand, integration of renewable energy resources and electric vehicles, transmission power losses, and improving the efficiency of the power system, are several challenges in modern power systems. Therefore, to overcome the mentioned challenges, macro and micro-grids are utilized to both enhance the power quality and increase the reliability of the grid side and the load side. There are several studies in the field of microgrids, considering their different manifestations, such as Smart Grids (SGs) and Virtual Power Plants (VPPs), etc. [11]. One of the main concerns in modern power system analysis is the dependency on the power from microgrids by using the grid power along with their developments. As a result, microgrids can interconnect to the power grid and improve the power quality and reliability. From another perspective, microgrids can connect to/disconnect from power grids to enable themselves and operate in both grid-connected and islanding modes.

Based on the expansion of the interconnected power grids through the long transmission lines, increasing the load demand, and the need for a supervisory control system for the power generation units, electric utilities are moving toward the decentralized and deregulated power systems, focusing on independent microgrids. Non-traditional power generation sources (e.g., wind farms, solar power plants, diesel generators, etc.) in microgrids are allowed to trade electricity with the local consumers. In addition, microgrids in a centralized structure no longer rely on a single power source. On-site generations can be utilized as emergency backups in the event of blackout or load shedding to mitigate disturbances and increase power system reliability. Figure 1 illustrates the concept of modern microgrids. As shown, there are several ways to utilize Distributed Generations (DGs), such as wind, solar, ESSs, etc. in the power system to support the grid and supply the load demand.

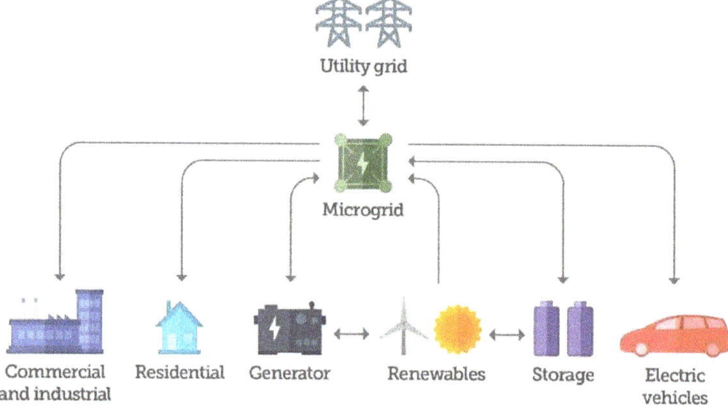

Figure 1. The concept of modern microgrids.

1.2. Literature Survey and Contributions

Technically, modern microgrids are a small part of low voltage distribution networks that are located far from the substation and interconnected through a Point of Common Coupling (PCC) [12]. Based on the nature of microgrids operation (e.g., ownership, location, reliability requirements, and trading purposes), significant developments are carried out by researchers, industrial and commercial factories, and military bases. According to the Power and Utilities Navigant Research, the capacity of the global microgrid has been projected to grow from 1.4 GW in 2015 to 7.6 GW in 2024 under a base scenario [12]. Modern microgrids not only offer great promises owing to their significant benefits but also result in tremendous technical challenges. There is an urgent need to investigate the state-of-art control and energy management systems in microgrids.

One of the main challenges in microgrids technology is to manage and balance the generation and consumption of energy [13]. The power imbalance is a typical scenario in microgrids, which comes from the nature and availability of renewable energy resources to discontinuously generate power and available loads connected to microgrids. The control system should manage these imbalances to prevent electrical damage and maintain AC/DC grids stable [14–17]. As a result, recent studies have focused on proper power management and control strategies to manage the generation and consumption of energy. These control strategies are mainly targeted at: (i) controlling the interconnected DGs and ESSs, (ii) DC bus voltage regulation, (iii) minimizing the cost of imported/exported energy from/to the main grid by optimizing the power dispatch between converters and DC bus voltage, (iv) management, and optimization of ESSs operation, and (v) current sharing management between parallel converters in DC grids [17–20].

In order to optimize the power dispatch, proper communication infrastructure between the microgrids and the grid operator are required [21,22]. However, real-time simulation and monitoring can be implemented by the communication infrastructure, the outage of the communication links/signals can cause many complicated problems. The droop control method is a well-known strategy to maintain the power balance in DC microgrids [17,22–26], which does not require communication infrastructure [17]. The power management of microgrids can be classified into centralized, decentralized, and distributed control categories [25–29]. The energy dispatched in the centralized control systems can be monitored and managed by an intelligent centralized (master) controller, which receives and analyzes the data, manages the power among the converter stations under operation, and forecasts the power and voltage references to all the power devices of the microgrids. [30,31]. These systems usually offer precise power-sharing among converters in microgrids [28,29]. In case of loss or outage of the master controller, local autonomous controllers (decentralized structure) are needed to fulfill the master controller failure [32–35]. In a distributed control system, each microgrid is allowed to only communicate with its nearby neighbors. Therefore, there is still a need for communication infrastructure. Further, there are many loops in a distributed control system, which make its design more complicated [36].

However, installing new components and/or upgrading the existing components are two methods to overcome the negative impacts of PHEVs in power grids, high investment costs prevent the mentioned solutions to be implemented. If the high penetration level of PHEVs is connected to the system, up to a 15% increase in the cost of upgrading the existing grid to guarantee the adequacy of the power system can be expected [37]. Therefore, more investigations are required to find a suitable and cost-effective solution. Utilizing proper ESSs and appropriate charging/discharging mechanism to control the power flow in PHEVs are preferable to installing new components and/or upgrading the existing infrastructures in power grids. High penetration chargers can be designed and implemented to allow bidirectional power flow between ESSs and power grids.

PHEVs can assist in improving the load-leveling profile and reducing power losses [38]. Further, utilizing efficient Voltage-Source Converters (VSCs) in power system allows transferring reactive power, as well as active power into the power grid. The DC-link capacitor and a proper switching mechanism can improve the quality of the transferred power into the grid [39]. Therefore, several studies have investigated different control strategies to implement the concept of a bidirectional charger and solve the charging issues of PHEVs [40–43].

A practical power electronics grid interface that can provide the Vehicle-to-Grid (V2G) bidirectional power flow with a high power quality is necessary to perform the grid-connected vehicle battery application. This interface should respond to the charge/discharge commands that are received from the monitoring system to enhance the reliability of the power grid. Moreover, essential requirements, such as reactive power injection and tracking the reference charge/discharge power, should be met [44,45]. A comprehensive review of the bidirectional converters is presented in [46], and discusses their advantages and disadvantages. In [47], the energy efficiency in PHEV charges along with the evaluation and comparison of the AC/DC topologies, such as the conventional Power Factor Correction (PFC) boost converter, including a diode-bridge rectifier followed by a boost converter, an interleaved PFC boost inverter, a bridgeless PFC boost converter, a phase-shifted semi bridges PFC boost converter [48], and a bridgeless interleaved boost converter [49] is presented. In addition, some non-inverted topologies are presented in [50–60], some of which require two or more switches to be operated in Pulse Width Modulation (PWM) mode, which causes higher total switching losses [50–52,54–61]. However, bidirectional power flow cannot be achieved in the topologies of [50,54,57,61–64].

Different DC charging station architectures for PHEVs are proposed in [65–67]. For instance, the control of the individual EV charging processes introduced in [65] is decentralized, while a separate central supervisory system controls the power transfer from the power grid to the DC link. With sufficient energy stored in the battery of PHEVs, the bidirectional charging/discharging power control of PHEVs can be applied to reduce the frequency fluctuation [68–74]. A power charging control system to control the frequency in the interconnected power system with wind farms is considered in [68]. The controller in [68] is capable of stabilizing the system frequency during the charging period. Further, the bidirectional power control of PHEV applied for frequency control in the interconnected power systems with wind farms is proposed in [69]. The proportional-based PHEV power controller in [69] provides satisfactorily control, but its performance may not tolerate such uncertainties, and it may fail to handle the system frequency fluctuation.

Further, the system parameters may not remain constant and continuously change when operating conditions vary [75]. The system parameter variations, such as the inertia constant and damping ratio, are conventionally considered to check the performance and robustness in the Load Frequency Control (LFC) approach [76]. Hence, the robustness of the controller against system uncertainties is a vital factor that must be considered.

This paper presents a precise bidirectional charging control strategy of PHEVs in power grids to simultaneously regulate the voltage and frequency, as well as reducing the peak load, and improving the power quality by considering the SoC and available active power in power grids. Different events that may occur during a 24-h scenario in the studied DG-based system consisting of different microgrids, diesel generator and wind farm, PHEVs with several charging profiles, and different loads are considered. The simulation and analysis are performed in MATLAB/SIMULINK software.

2. Principles of Bidirectional Power Flow

Figure 2 shows a basic model of a power system consisting of two generators and a transmission line which is connected to two generation buses. In this figure, each bus has its voltage magnitude and phase angle, where $V_1\angle\theta_1$ and $V_2\angle\theta_2$ are the corresponding voltage magnitude and phase angle of buses 1 and 2, respectively. The impedance of the transmission line is $Z\angle\gamma$, where:

$$Z = R + jX \tag{1}$$

and

$$\gamma = \tan^{-1}\frac{X}{R} \tag{2}$$

Therefore, by considering voltage magnitude and phase angle differences between buses, the active power, P, and reactive power, Q, can be transferred, bidirectionally. In order to study the bidirectional power flow, it should be assumed that both buses are capable of supplying and absorbing active and reactive power.

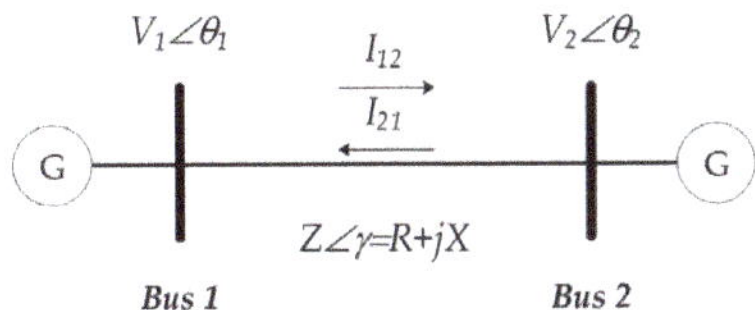

Figure 2. Single diagram of a two-bus system.

Based on the direction of power flow, the active and reactive power equations can be written as:

$$P_{12} = \frac{|V_1^2|}{|Z|}\cos\gamma - \frac{|V_1||V_2|}{|Z|}\cos(\gamma + \theta_1 - \theta_2) \tag{3}$$

$$Q_{12} = \frac{|V_1^2|}{|Z|}\sin\gamma - \frac{|V_1||V_2|}{|Z|}\sin(\gamma + \theta_1 - \theta_2) \tag{4}$$

Assuming $X \gg R$, Equations (3) and (4) can be re-written as:

$$P_{12} = \frac{|V_1||V_2|}{X}\sin(\theta_1 - \theta_2) \tag{5}$$

$$Q_{12} = \frac{|V_1|}{X}[|V_1| - |V_2|\cos(\theta_1 - \theta_2)] \tag{6}$$

Small changes in the voltage magnitude have a direct impact on the reactive power flow, while deviations in phase angle can change the active power flow in power grids. If $\theta_1 > \theta_2$, the active power can be transferred from bus 1 to bus 2 and vice versa. Further, if $V_1 > V_2$, the reactive power can be transferred from bus 1 to bus 2 and vice versa. Therefore, the voltage magnitude and phase angle play the important roles in power transfer in power grids.

3. Control Strategy and Power System Modeling

3.1. Bidirectional Charging Station

PHEV chargers should be installed off-board and onboard in a vehicle. Onboard chargers are built with a small size, low power rating, and can be used based on a slow charging mechanism. Off-board PHEV chargers are located at specific places and provide either a slow or fast charging mechanism. As a result, charging networks play an important role to support PHEVs. In addition, there are two common architectures (series and parallel) in the PHEV drivetrain. Moreover, a combination of these two (series-parallel architecture) is also used in some vehicles [1,5–8,77]. This paper has considered a generic aggregation of PHEVs with different charging profiles. The number of vehicles in charge, the rated power and rated capacity, and the power converter efficiency are the important factors in this model. This model is also capable of enabling vehicles to the grid, instantly. Table 1 shows the charging station specifications.

Table 1. Charging station specifications.

Parameter	Descriptions
Rated Power	40 kW
Rated Capacity	85 kWh
Rated Voltage	50–600 V_{DC}
System Efficiency	90%
No. of cars (different profile)	35, 25, 10, 20, 10

The fundamental configuration of the charging station consists of a centralized AC/DC converter and different DC/DC converters as the charging lots. The AC/DC converter rectifies the three-phase AC input signal into a DC output signal. DC/DC converters have been used to regulate and shift the output signal to the desired level. A bidirectional charging station is capable of controlling the power flow in both directions between charging stations and power grids.

Figure 3 illustrates the single-line diagram of the case study in this paper [78]. Further, the proposed scheme of the grid-connected PHEV system is illustrated in this figure. The proposed model has a central AC/DC VSC station and different controllable DC/DC converters based on a certain number of PHEVs. All DC/DC converters are in a parallel architecture and are connected to a common DC bus, which has been regulated by the central AC/DC VSC station. Two different conditions have been assumed for the charging stations: (1) cars in regulation, which contribute to the grid regulation; and (2) cars in charge, which are regularly in the charging process. Therefore, there are two modes of operation for each PHEV: (1) regulation mode; and (2) charging mode.

In order to minimize the output harmonics during the operation and switching processes of converters, an RL filter has been considered between the charging station and power grid. The charging system in this paper allows transferring the active and reactive power bidirectionally with the help of a control system. The control strategy mainly focuses on the AC/DC and DC/DC converters. By proper controlling the central AC/DC VSC station, injecting reactive power into the power grid to regulate the voltage, improving the power factor, and maintaining a constant DC-bus voltage, can be achieved. Moreover, an appropriate control mechanism for the DC/DC converters can ensure controlling the charging and discharging processes of PHEVs. This paper has considered both charging and discharging operations, where the DC/DC converters are controlled by charge/discharge PHEV batteries.

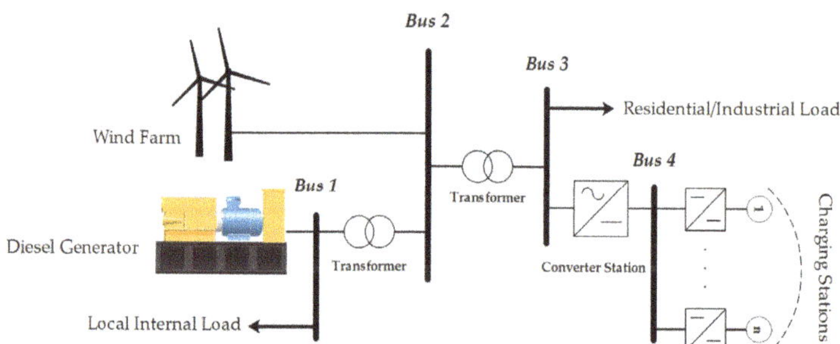

Figure 3. Single line scheme of the modified microgrid system (case study).

3.2. Converter Station Control Systems

As stated in Section 3.1, the converter station can work in two modes: (1) the regulation mode, and (2) the charge mode, and is capable of compensating reactive power, and consequently, regulating the voltage of the power grid during the charging and discharging processes of PHEVs. This section provides details of the control systems of the converter station in different modes.

3.2.1. Grid Regulation Mode

In order to contribute to the grid regulation, two controllers have been designed: (1) a grid regulation controller, and (2) a grid regulation power generation. Figure 4 shows the grid regulation controller scheme. This controller is fed by the maximum regulated active power, P_{max_reg}.

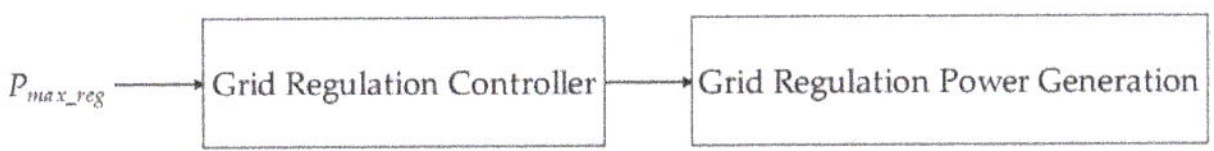

Figure 4. Grid regulation controller scheme.

To achieve the maximum active power regulation, a control system that consistently checks the nominal active power, P_{nom}, is proposed, as shown in Figure 5. In this controller, the nominal active power, the number of cars in regulation, N, and an online key to enable the V2G mode are matter. Lastly, the output power has been limited within the standard range to be considered as the maximum regulated active power. There is a threshold (0.5) for passing the first input to the second input, C_1, of this controller.

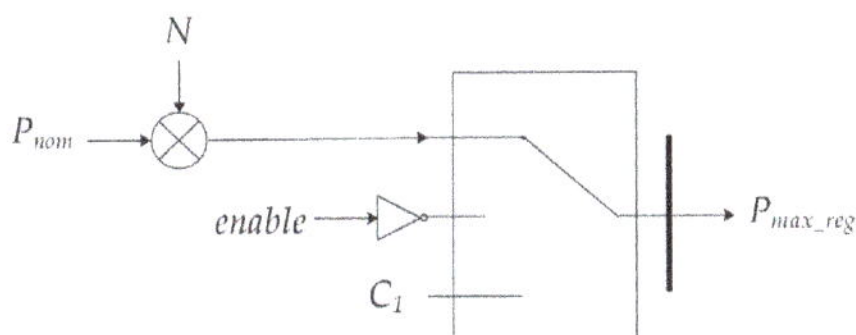

Figure 5. Outer control system scheme for gaining maximum regulated active power.

In order to contribute to the grid regulation, real-time measurement of the frequency, ω_{grid}, is required. By comparing ω_{grid} and the reference frequency, ω_{ref}, the frequency deviation can be obtained. The frequency deviation should be less than 0.05%. Otherwise, the controller stops the process. To prevent sudden changes in the frequency deviation, the controller's derivative has been utilized. By considering that the frequency deviation is within the standard range, two gains for the grid regulation controller have been set by the operator, the open-loop gain, K_1, and the loop gain, K_2. Changing these two gains has a direct impact on the SoC of all the cars in the charge mode. A zero-crossing detection integrator has been considered to minimize the disturbance and steady-state error of the input signal. The output signal of the integrator has been rechecked to avoid the maximum allowable regulated active power that is fed into the grid regulation power generation system. Figure 6 illustrates a detailed diagram of the grid regulation controller.

Furthermore, by controlling the voltage, and consequently, the current through another control system, the proposed control strategy can ensure a contribution to the grid regulation power generation, as shown in Figure 7.

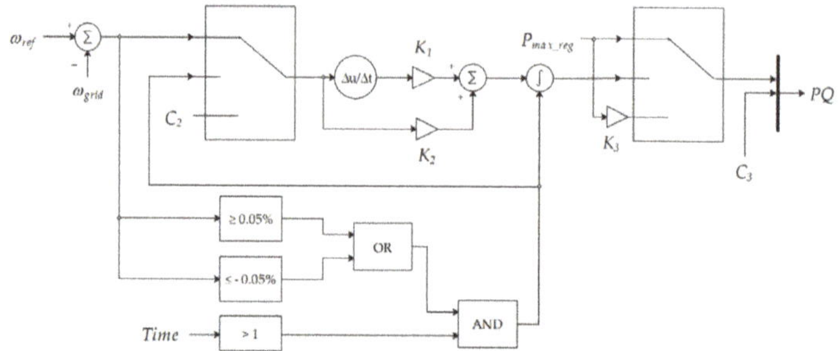

Figure 6. Control diagram of the grid regulation controller.

The grid regulation power generation controller captures the voltage of pair phases. Therefore:

$$V = \frac{1}{3}\left(V_{ab} - a^2 V_{bc}\right) \tag{7}$$

where V_{ab} is the voltage between phases a and b, and V_{bc} is the voltage between phases b and c.

Moreover:

$$a^2 = e^{-j\frac{2\pi}{3}} \tag{8}$$

By using Equation (7) and decomposing the real and imaginary parts from the apparent power, S, the current can be derived as follows:

$$I = \frac{2}{3}\frac{S^*}{V^*} \tag{9}$$

where V and I denote the voltage and current, respectively. S^* and V^* represent the complex conjugate of the apparent power and the complex conjugate of the voltage, respectively. Accordingly, the controller feeds constant current into the power grid and regulates the voltage and frequency, simultaneously.

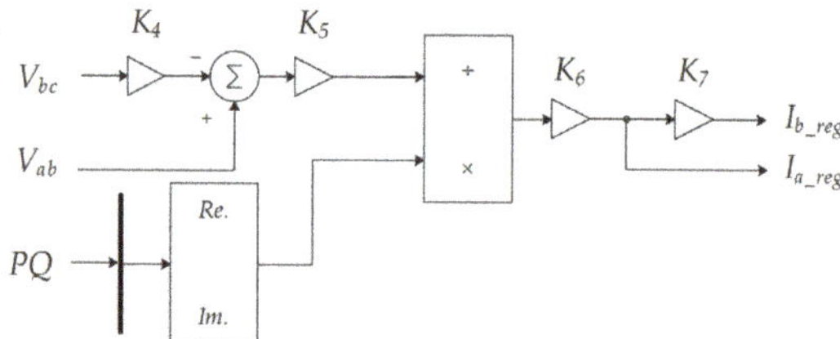

Figure 7. Control diagram of the grid regulation power generation.

3.2.2. Charge Mode

In order to control the PHEV station in charge mode, two states have been studied: (1) SoC, and (2) plug. The charge power generation controller requires the nominal active power and the number of cars in charge, M, as its inputs. Same as the grid regulation mode, a threshold (0.5) is used for passing the first input to the second input, C_4, of the charge power generation control system. The threshold can guarantee safe operation. Figure 8 shows the outer control diagram of the charge power generation controller.

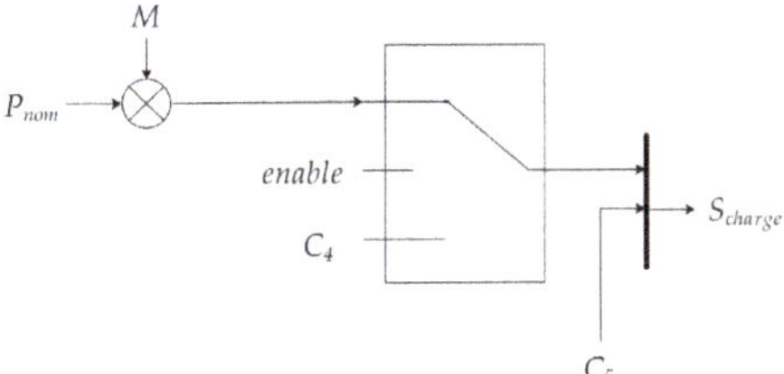

Figure 8. The outer control diagram of the charge power generation controller.

Same as the previous section, by regulating the voltage, and decomposing the real and imaginary parts of the apparent power of the charging system, a constant current, I_{reg}, is obtained from the controller, and fed into the power grid. Figure 9 illustrates the inner control diagram of the charge power generation controller.

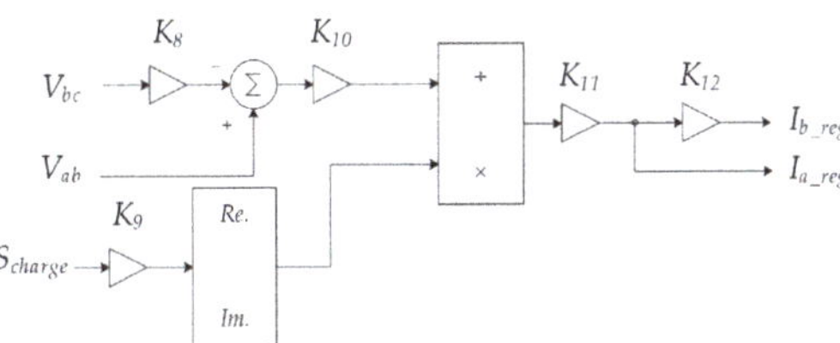

Figure 9. The inner control diagram of the charge power generation controller.

Each group of cars has a certain charging profile. As mentioned, the car profile has been investigated by checking the SoC and plug states. Figure 10 demonstrates the control diagram of the profile of each PHEV in charge and regulation modes. In this control design, the SoC initialization and plug state have been implemented by using the Binary Search Method (BSM). Therefore, the SoC initialization and plug state have stochastic, but linear behavior. Complementary descriptions for the different profiles are provided in Section 3.7.

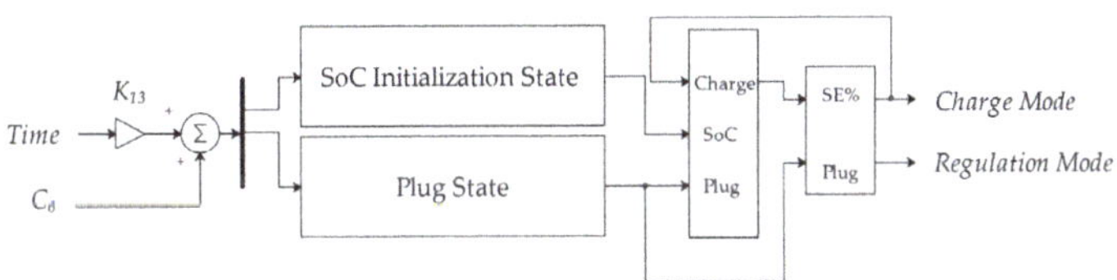

Figure 10. The control diagram of the profile of each PHEV in charge and regulation modes.

In order to control the output changes of the SoC initialization and plug state within limits, two state limiters have been considered. By capturing the SoC initialization and plug state, and the present values of the SoC in different profiles, the state estimator for each SoC controller has been designed. Figure 11 indicates the SoC controller, where the output of this controller is the State Estimator (*SE*%), and regulated for the charger controller. The SoC controller needs accurate information of the active power of the cars in charge, P_{charge}, the number of cars in charge mode, M, the number of cars in the specific charging profile, L_i (where $i = 1, \ldots, 5$), the number of cars in regulation mode, N, the active power of the cars in EV mode, P_{EV}, and the regulated output of the SoC initialization and plug state.

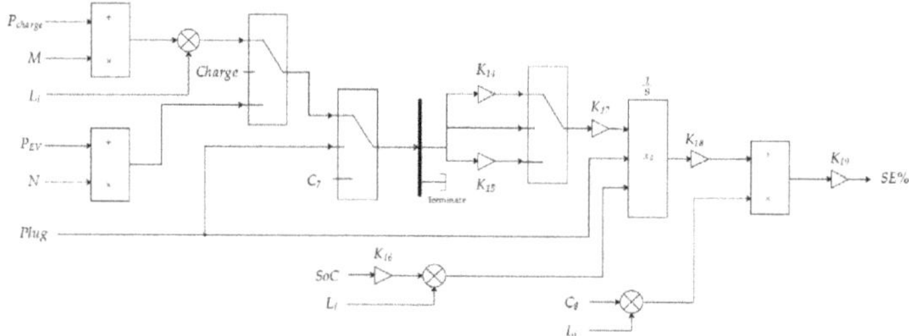

Figure 11. The control diagram of the SoC controller.

The state estimation of batteries in PHEVs requires checking the mode of vehicles. The PHEV can be either in charge or EV mode. Therefore, there should be a switch to toggle between these two modes. By reaching the maximum level of the charge, the controller terminates the charging process. In the meantime, the controller checks the plug state dynamically, and if the PHEV is unplugged from the grid, it sends zero signal, C_6, as shown in Figure 11, and terminates the charging process.

Assuming the PHEV is in charge mode, the charge/discharge efficiency has been taken into consideration. These two are modeled as two direct gains, K_{14} and K_{15}, so that:

$$\begin{cases} K_{14} = \eta \\ K_{15} = 1/\eta \end{cases} \tag{10}$$

where η shows the efficiency.

The SoC, which is converted to real-time by multiplying the present capacity by 1000 × 3600, the plug state, and the charging state are sent to an integrator with the initial condition and dynamic saturation, and lastly, the output is multiplied by the number of cars in a particular profile to obtain the state estimation. State estimation from this stage is used as the SoC of the vehicles. Deriving the state estimation, the cars are checked whether they are either in the charge mode or regulation mode through the charger controller. Figure 12 indicates the charger control diagram. In order to guarantee a high-efficiency output during the charging process, the state estimation has been set within the range of 85% and 95%. Otherwise, the charger terminates the process. In fact, this can ensure both injecting power with high quality to the power grid during the regulation mode and charging the batteries during the charging mode.

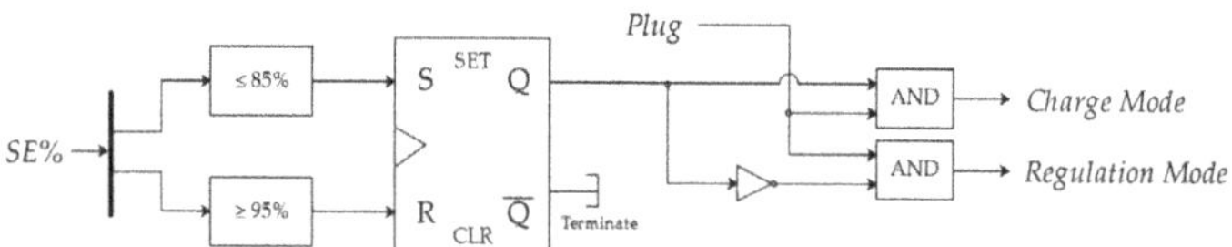

Figure 12. Control diagram of the charger.

Proper operation of the charging station of PHEVs lead to the energy balance of ESSs in PHEVs subject to the maximum and minimum operating limitations in charging and discharging power as follows:

$$E_{s,i,(t+1)} = E_{s,i,t} + \eta_{s,i,c} \times P_{s,i,t,c} - \frac{P_{s,i,t,disc}}{\eta_{s,i,disc}} \tag{11}$$

subject to:

$$0 \le P_{s,i,t,c} \le k_{s,i,t,c} \times P_{s,i,c\_max} \tag{12}$$

$$0 \le P_{s,i,t,disc} \le k_{s,i,t,disc} \times P_{s,i,disc\_max} \tag{13}$$

$$k_{s,i,t,c} + k_{s,i,t,disc} \le 1 \tag{14}$$

$$k_{s,i,t,c} + k_{s,i,t,disc} \in \{0, 1\} \tag{15}$$

$$E_{s,i,min} \leq E_{s,t} \leq E_{s,i,max} \tag{16}$$

where indices s, i, and t refer to the sth energy storage system at the ith bus in the tth time interval. Therefore, $E_{s,i,t}$ is the energy storage of the sth energy storage system at the ith bus in the tth time interval in MWh. $\eta_{s,i,c}$ and $\eta_{s,i,disc}$ are charging and discharging efficiencies, respectively. $P_{s,i,t,c}$ and $P_{s,i,t,disc}$ are charge and discharge power, and $k_{s,i,t,c}$ and $k_{s,i,t,disc}$ are the binary variables for charging and discharging operations of the sth energy storage system at the ith bus in the tth time interval, respectively.

Figure 13 shows the flow chart of the proposed bidirectional power charging strategy. According to the collected data from the installed meters, the amount of the active and reactive power in the system is checked. When V2G is not activated, the existing microgrid(s) satisfy the total load consumption, whether there is a contingency in the system or not. When V2G is activated, the SoC initialization and plug state of PHEVs are checked and based on them, the SoC can be estimated. According to the estimated SoC, PHEVs mode can be either in charge or regulations mode. By considering that there is no contingency in the system, the required power to charge PHEVs is supplied by the existing microgrid(s). Otherwise, all PHEVs in charge and regulation modes are passed through a regulation unit, and the regulated voltage and frequency are then used to update the active and reactive power.

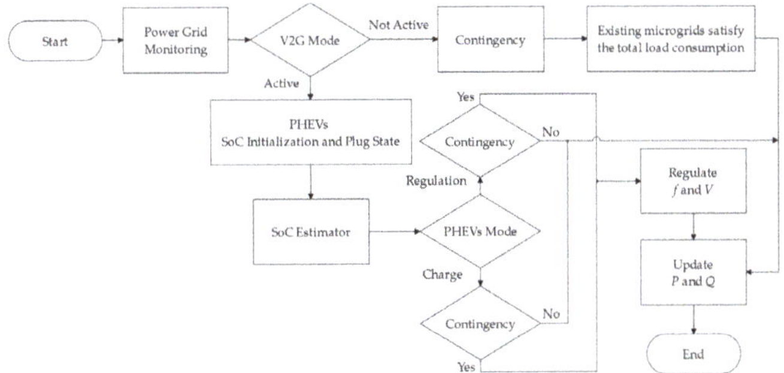

Figure 13. Flow chart of the proposed bidirectional power charging strategy.

3.3. Diesel Generator

A three-phase synchronous machine in the dq-rotor reference frame with the engine governor and excitation system has been modeled and considered as the diesel generator to feed 15 MW active power to the power grid. The nominal power, line-to-line voltage, and frequency of the generator are 15 MW, 25 kV, and 60 Hz, respectively. The IEEE type 1 synchronous generator voltage regulator combined with an exciter has been implemented as the excitation system. Moreover, the diesel governor has been modeled, where the desired and actual rotor speeds are the inputs and the mechanical power of the diesel engine is the output. Further, the motor inertia has been combined with the generator. The design considerations for the governor are made through the regulation of the controller and actuator as follows:

$$H_c = \frac{K(1 + T_3 s)}{(1 + T_1 s + T_1 T_2 s)} \tag{17}$$

where H_c is the controller transfer function, K is the regulator gain, and T_1, T_2, and T_3 are the regulator time constants, respectively.

Furthermore:

$$H_a = \frac{(1 + T_4 s)}{s((1 + T_5 s)(1 + T_6 s))} \tag{18}$$

where H_a is the actuator transfer function, and T_4, T_5, and T_6 are the regulator time constants, respectively. The engine time delay, T_d, has been set to 0.024 sec. It should be noted that ω_{ref} and V_{ref} of the diesel engine governor and excitation system have been set to 1 p.u.

Accordingly:

$$T_m - T_e - D\Delta\omega = J\frac{d\omega}{dt} \tag{19}$$

where $T_m = \frac{P_m}{\omega_m}$, $T_e = \frac{P_e}{\omega_e}$, and $\Delta\omega = \omega - \omega_{rated}$. T_m, T_e, ω_m, and ω_e are the mechanical torque of the synchronous generator rotor shaft input from the prime mover, stator electromagnetic torque, mechanical speed of the rotor, and synchronous speed, respectively. D indicates the damping coefficient, ω and ω_{rated} are the actual electrical and rated angular velocities, respectively. P_m and P_e are the mechanical and electromagnetic power, respectively. J shows the rotational inertia, and θ indicates the electrical angle.

In addition:

$$E_0 = V_s - I(r_a - jx_s) \tag{20}$$

where E_0, V_s, and I are the three-phase stator winding electromotive force, the stator voltage and current, respectively. Moreover, r_a and x_s indicate the armature resistance and stator reactance, respectively.

3.4. Wind Farm

As the second source of energy, a wind farm with 4.5 MW nominal power capacity, with 13.5 m/s as the nominal wind speed, and 15 m/s as the maximum wind speed, is studied in this paper. The wind speed fluctuations provide a situation to study the power grid under a stochastic condition (uncertainty). The generated power by the wind farm can be obtained as follows:

$$P_{wind} = \begin{cases} 0 & 0 \leq v_i < v_{cut-in} \\ \frac{v_i - v_{cut-in}}{v_r - v_{cut-in}} \times P_r & v_{cut-in} \leq v_i < v_r \\ P_r & v_r \leq v_i < v_{cut-out} \\ 0 & v_r \geq v_{cut-out} \end{cases} \tag{21}$$

where P_{wind} is the active power generated by the wind farm, P_r shows the rated power of the wind farm, v_i indicates the wind speed, and v_{cut-in}, v_r, and $v_{cut-out}$ are the cut-in, rated, and cut-out speeds of the wind turbine, respectively. The output power of the wind farm has been treated as a negative load so that the power factor can be kept at a constant level.

3.5. Loads

A three-phase squirrel-cage asynchronous machine 1.5 MVA, 600 V, 60 Hz in the *dq*-reference frame (as the industrial load), and a set of loads with the nominal power of 10 MW at 0.95 power factor (as the residential load) have been modeled as the loads' case study. This model can represent the impact of inductive loads on the microgrid. The residential load follows a specific profile with the assigned power factor during the day. Further, the industrial load has been controlled by the square relation between the mechanical torque and the rotor speed.

3.6. Power Transformers

Two three-phase power transformers, a 20 MVA, 25kV/25kV, 60Hz transformer for the voltage regulation with Yg-Yg winding connections and a 20 MVA, 25kV/600V, 60 Hz step-down transformer with Yg-Yg winding connections have been modeled as a part of transmission systems for the power system.

3.7. *Power System Modeling*

As shown in Figure 3, a single-line diagram of a power system consisting of different microgrids and PHEVs has been modeled, where the total peak load is 10.15 MW, and the total generated power is 19.5 MW. One of the contributions of this paper is to consider variable power loading levels for PHEVs, which lead to different profiles for the charging stations. Five different profiles have been assigned to PHEVs. Profile 1 is for vehicles with the possibility to be charged during the working hours. Profile 2 is for vehicles with the possibility to be charged during working hours but with a longer ride. Profile 3 is for vehicles with no possibility to be charged during the working hours. Profile 4 is for vehicles which are parked at home and charged during the whole day. Profile 5 is for vehicles that are charged during the night shift. In this case, the impact of charging on the power grid has been investigated. In order to differentiate between the SoC of the PHEVs and the stand-alone battery packs, different charging and discharging cycles have been studied in this paper. The generator balances the power and load demand. The generator determines the frequency deviations of the grid at the rotor speed. By using transformers, the voltage level has been stepped-down to suitable voltage levels for the power grid. Table 2 shows the corresponding values of the control parameters for the charging station system in the power grid.

Table 2. Control parameters of the charging station.

Parameter	Value
$C_1, \ldots, C_5$ and C_7	0
C_6	1
C_8	85×10^3
K_1	2
K_2	4×10^3
K_3, K_9, and K_{17}	-1
K_4, K_7, K_8, and K_{12}	e^(-j2π/3)
K_5 and K_{10}	1/3
K_6 and K_{11}	2/3
K_{13} and K_{18}	1/3600
K_{14}	1/(90%)
K_{15}	90%
K_{16}	306×10^6
K_{19}	100

It is well-known that the PHEV charging process highly depends upon the connection point to the power system. This means that by connecting the PHEV to a weak bus, more power drains from the grid and the voltage drop increases, and consequently, it has a negative impact on the power grid.

Two scenarios have been investigated in 24 h. The wind speed varies during the day and has multiple maximum and minimum values. The residential load consumption profile is similar to that of the real world. The demand is low during the day, increases to the peak value during the evening and night, and gradually decreases during the late night. Two events have significantly affected the grid frequency during the day: (1) event 1, which is the asynchronous machine (industrial load) start-up at t = *03:00 a.m.*, and (2) event 2, which is the wind farm outage at t = *10:00 p.m.*, when the wind speed exceeds the maximum speed. The case study has been simulated under two different conditions for vehicles in regulation and charging modes.

4. Results and Discussions

The case study in this paper conducts the power profile (the generated and consumed power) as the bidirectional power flow during a 24-h. Contributions of the diesel generator and wind farm and the impact of PHEVs on the peak load reduction have been studied in this section. To study the bidirectional power flow, the active and reactive power balance of the system have been determined as follows:

$$P_{g_{i,t}} + P_{w_{i,t}} + P_{L_{i,t}} - P_{s,i,c} + P_{s,i,disc} - \sum_j [\frac{V_{i,t}V_{i,t}}{Z_{ij}} \cos\theta_{ij} + \frac{V_{i,t}V_{j,t}}{Z_{ij}} \cos(\theta_i - \theta_j + \theta_{ij})] = 0 \quad (22)$$

$$Q_{g_{i,t}} - Q_{L_{i,t}} - \sum_j [\frac{V_{i,t}V_{i,t}}{Z_{ij}} \sin\theta_{ij} + \frac{V_{i,t}V_{j,t}}{Z_{ij}} \sin(\theta_i - \theta_j + \theta_{ij})] = 0 \quad (23)$$

subject to:

$$\frac{V_{i,t}V_{j,t}}{Z_{ij}} \cos(\theta_{i,t} - \theta_{j,t} + \theta_{ij,t}) - \frac{V_{i,t}V_{j,t}}{Z_{ij}} \cos\theta_{ij,t} \leq P_{ij,max} \quad (24)$$

$$\frac{V_{i,t}V_{j,t}}{Z_{ij}} \sin(\theta_{i,t} - \theta_{j,t} + \theta_{ij,t}) - \frac{V_{i,t}V_{j,t}}{Z_{ij}} \sin\theta_{ij,t} \leq Q_{ij,max} \quad (25)$$

where indices i, j, s, and t refer to the bus i, bus j, the sth energy storage system, and the tth time interval. $P_{g_{i,t}}$, $P_{w_{i,t}}$, $P_{L_{i,t}}$, $Q_{g_{i,t}}$ and $Q_{L_{i,t}}$ show the active power generated by the non-renewable energy source, the active power generated by the wind farm, the active power consumed by the load, the reactive power generated by the non-renewable energy source, and the reactive power consumed by the load at the ith bus in the tth time interval, respectively. Further, V, θ, and Z indicate the voltage magnitude, and angle, and the impedance of the bus, respectively. $P_{ij,max}$ and $Q_{ij,max}$ are the maximum allowable active and reactive power that can be transferred between the buses, respectively.

4.1. V2G Mode is Deactivated

Based on the results of the simulation for *86,400 sec.* (24 h), which is shown in Figure 14, when the V2G system is not under operation, due to the defined scenarios during two different time intervals, one at t = *03:00 a.m.* (*10,800 sec.*) and the second one at t = *10:00 p.m.* (*79,200 sec.*), the industrial load bus voltage shows a significant change. Due to the first event, the voltage at all buses changes. Based on the nature of the load profile, the industrial load is not under operation before the third hour (the output power is 0 MVA), and the residential load reaches its minimum value (5.446 MVA), because of less usage of the normal resistive and inductive loads, such as lightings, refrigerators, etc. Therefore, the amount of current, and consequently, the drained power from the grid is not significant. At t = *10,800 sec.*, the industrial load starts up, and the power flow in the grid changes. The voltage at the industrial load bus drops, and the current increases drastically (from 0 to 2184 A). Hence, 445.3 kW power is extracted as the power losses due to this event, and accordingly, this power is supplied by the diesel generator and wind farm. The total load demand at that time reaches 7.235 MVA, and the power grid supplies the load through the amount of generated power by the two microgrids, where the generation contributions of the diesel generator and wind farm are 5.328 MVA and 2.446 MVA, respectively. Thus, the total generated power is derived as 7.774 MVA, which is more than the total load demand. It should be noted that when the V2G mode is deactivated, by increasing or decreasing the corresponding values of the controller's gains (K_1 and K_2), the SoC of the batteries of PHEVs in the charge mode does not change.

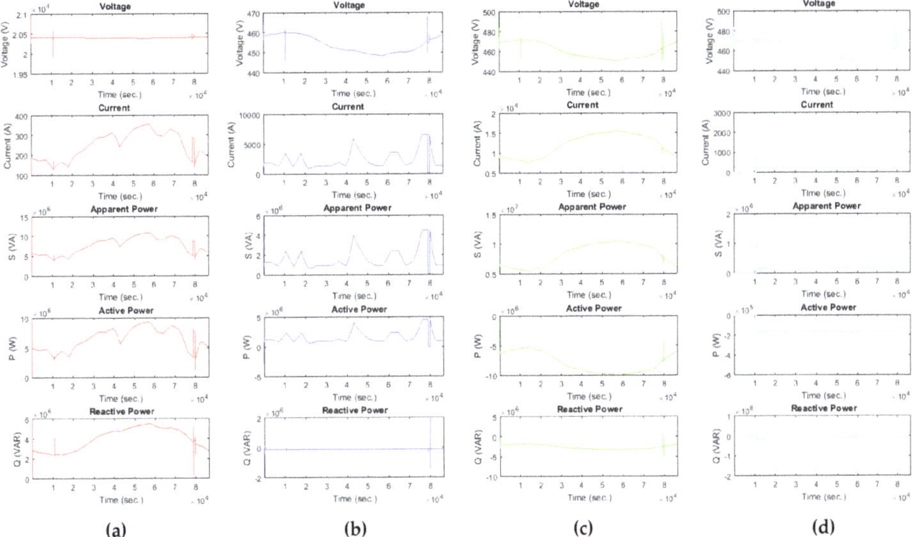

Figure 14. Voltage, current, the apparent power, active power and reactive power curves of (**a**) the diesel generator, (**b**) wind farm, (**c**) residential load, and (**d**) industrial load during 24 h when the V2G system is not under operation.

As shown in Figure 15, the total load consumption increases from 6.0980 MVA at *05:00 a.m.* *(21,600 sec.)* to 8.1800 MVA at *10:00 p.m.* *(79,200 sec.)*. Based on the results of the simulation, the diesel generator and wind farm successfully supply the load demand, even during the contingency. The transient time related to the voltage regulation at each bus is minimized by proper operating of the controllers. This is achieved by defining the power flow constraints (Equations 24 and 25) for the power grid. When the wind farm has less generation or is not under operation for a certain period of the time, the diesel generator acts as a fast-response power supply to supply the load without interruption. Table 3 shows the power flow of the power grid during the entire day.

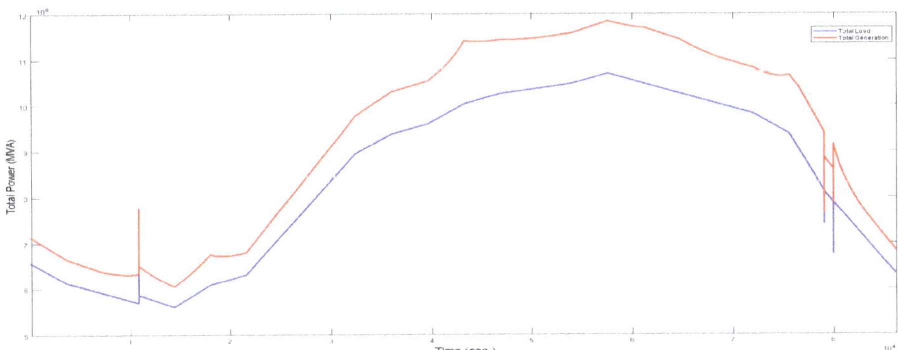

Figure 15. Total generated and consumed power curves in 24 h when the V2G system is not under operation.

Table 3. Power flow results when the V2G system is not under operation.

Time	Diesel / Wind	Total Generation	Total Demand	Power Losses
00:00 a.m.	3.3660 0.7000	4.0660	3.8480	0.2180
01:00 a.m.	5.3350 1.3300	6.6650	6.1430	0.5220
02:00 a.m.	5.4550 0.9340	6.3890	5.9210	0.4680
03:00 a.m.	5.3280 2.4460	7.7740	7.2350	0.5390
04:00 a.m.	5.1270 0.9340	6.0610	5.6160	0.4450
05.00 a.m.	4.3380 2.4260	6.7640	6.0980	0.6660
06:00 a.m.	6.1750 0.6270	6.8025	6.3170	0.4855
07:00 a.m.	6.8530 0.9330	7.7860	7.1920	0.5940
08:00 a.m.	7.8370 0.9340	8.7710	8.0710	0.7000
09:00 a.m.	8.8380 0.9340	9.7720	8.9470	0.8250
10:00 a.m.	8.9790 1.3290	10.308	9.3830	0.9250
11:00 a.m.	9.6030 0.9350	10.538	9.6060	0.9320
12.00 p.m.	7.4050 4.0110	11.416	10.004	1.4120
01:00 p.m.	9.0070 2.4300	11.437	10.260	1.1770
02:00 p.m.	10.140 1.3300	11.470	10.370	1.1000
03:00 p.m.	10.640 0.9340	11.574	10.480	1.0940
04:00 p.m.	10.910 0.9340	11.844	10.700	1.1440
05.00 p.m.	9.2090 2.4290	11.638	10.430	1.2080
06:00 p.m.	9.0070 2.4290	11.436	10.260	1.1760
07:00 p.m.	10.120 0.9340	11.054	10.040	1.0140
08:00 p.m.	9.4940 1.3290	10.823	9.8260	0.9970
09:00 p.m.	6.6540 4.0100	10.664	9.3870	1.2770
10:00 p.m.	8.3160 0.0120	8.3250	8.1800	0.1480
11:00 p.m.	6.8520 0.9340	7.7860	7.1950	0.5910

Note: All the values are in MVA and have been rounded to the closest number.

Figure 16 shows the dynamic behavior of the converter station during the simulation time. As long as the breaker of the converter station is open and the V2G system is not under operation, the power consumption by the converter station is completely insignificant. However, the station detects the two events during the simulation time. It should be noted that the consumed power by the converter station is considered as the power losses in Table 3. As shown in Figure 16, in the regulation and charge modes, the consumed power by the converter station is close to zero.

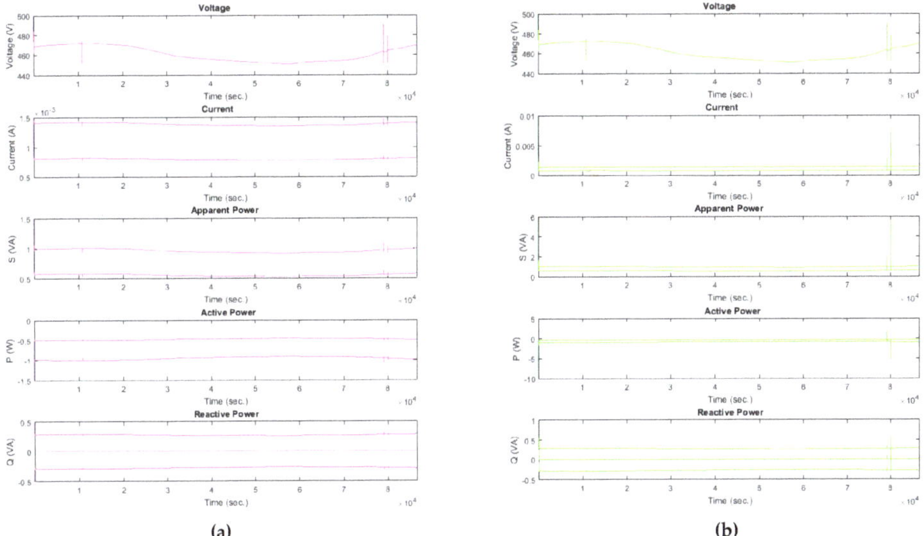

Figure 16. Voltage, current, the apparent power, active power, and reactive power curves of the converter station in (**a**) the charge mode and (**b**) the regulation mode in 24 h when the V2G system is not under operation.

4.2. V2G Mode is Activated

When the V2G mode is activated, the operating impact of the V2G system changes the power flow of the grid. Theoretically, it is expected that when the V2G system is under operation, more power is supposed to be drained from the power grid. The more power consumed by the converter station, the greater the power required from the diesel generator and wind farm. Unlike the previous mode that the V2G mode was deactivated, the different car profiles change the total load profile. Therefore, the peak load is expected to be more than the previous operating mode. Figure 17 shows the results of the simulation in the V2G system operation.

Unexpected events (contingency and/or outage) in the power system lead to a change in the voltage and frequency. Heavy load or generator outages can be considered as such changes that influence the voltage and frequency variations. As shown in Figure 18, based on the two defined scenarios, at t = *10,800 sec.*, the industrial load starts up and drains power from the generation buses. The total generated power should satisfy the total load demand, including the cars in profiles 4 and 5. Due to the sudden changes in the power, the converter station detects the voltage and frequency variations, and the controllers switch to the regulation mode and contribute to the grid regulation, as shown in Figure 18. However, the voltage curve in both charge and regulation modes fluctuates around its nominal value, the frequency deviations related to the power changes after the contingency are more visible. Due to rapid fluctuations in the voltage of the power supply or loads, a momentary flicker can also be observed in the power system. The designed system attempts to mitigate and eliminate this momentary flicker by regulating (or stabilizing) the voltage.

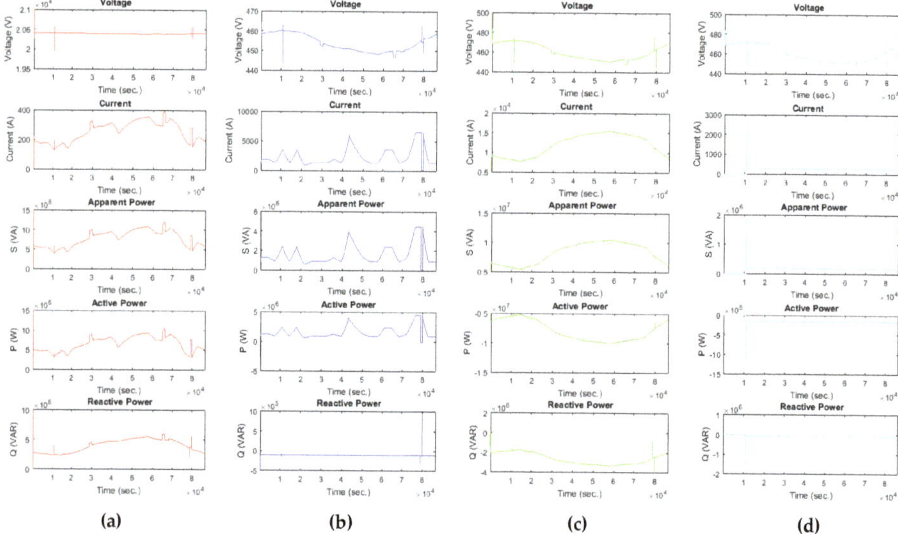

Figure 17. Voltage, current, the apparent power, active power, and reactive power curves of (**a**) the diesel generator, (**b**) wind farm, (**c**) residential load, and (**d**) industrial load during 24 h when the V2G system is under operation.

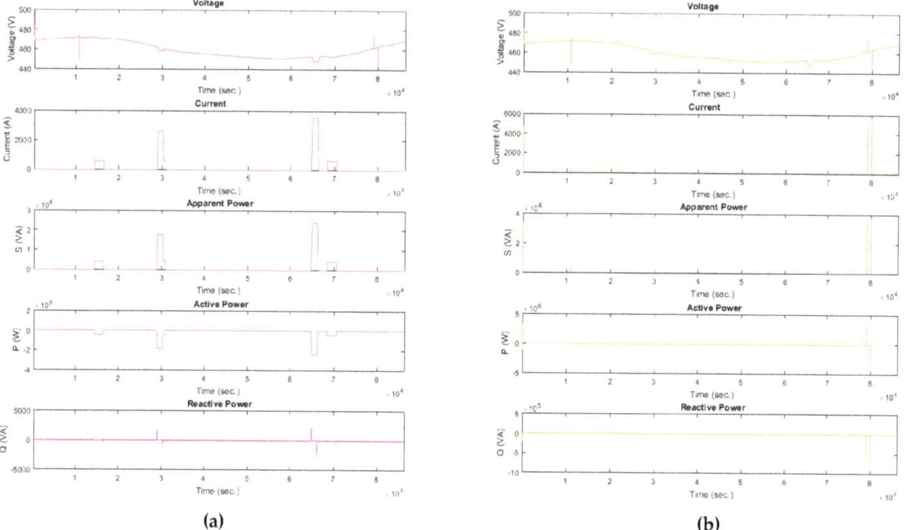

Figure 18. Voltage, current, the apparent power, active power, and reactive power curves of the converter station in (**a**) the charge mode and (**b**) the regulation mode in 24 h when the V2G system is under operation.

According to the descriptions of the car profiles in Section 3.7, the behavior of the SoC varies based on the car profile. Figure 19 illustrates the SoC of the different car profiles during the day, when the V2G system is under operation. When the second scenario occurs at *t = 79,200 sec.*, the level of the generated power is reduced significantly. Consequently, PHEVs switch to the regulation mode to contribute to the voltage and frequency regulation and restore the system to its previous condition. Because the cars in profile 4 are connected to the grid for the entire day, their contributions to the grid regulation are more severe. Also, due to the grid mode controller settings, a gradual increase in the SoC is expected. Due to the fact that the wind farm outage decreases the level of generated power, the system can reach a critical condition without utilizing proper control systems.

It is observed that when the V2G mode is activated, by decreasing K_1, the SoC of the batteries decreases very slightly during the simulation time. By increasing K_1, the SoC of the cars in profile 4 increases from 90.0 to 90.6. Decreasing K_2 causes an inverse trend in the SoC variations. This means that the SoC of the cars in profile 4 increases, and then, gradually decreases. By increasing K_2, the SoC of the batteries of the cars in profile 4 moderately decreases during the simulation time. It should be noted that the minimum level of the SoC is 10%.

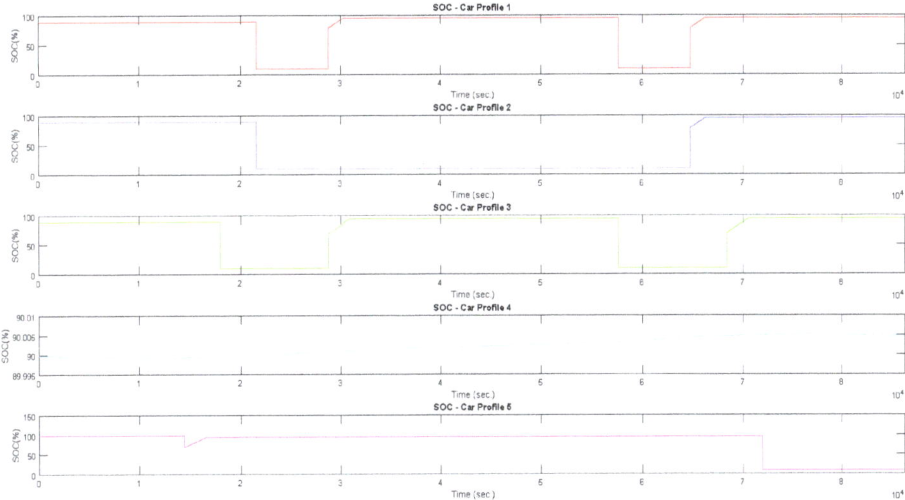

Figure 19. SoC of the different car profiles in 24 h when the V2G system is under operation.

Figure 20 illustrates the total generated and consumed power curves when the V2G system is under operation. As shown in this figure, the total load demand varies throughout the day and is met by the dispatchable generation unit. The generation units adjust operations to follow the load pattern. When the V2G system is not under operation, there is only one peak point during the entire simulation time. However, when the V2G mode is activated, the number of peak points is increased due to the different car profiles. The peak points are met when the number of car-users is increased, and the generation units are forced to generate more power. The maximum peak point has been detected between *t* = *17:00* and *t* = *18:00*, because cars in profiles 1, 3, 4, and 5 are all in the charging mode. Although the wind farm has been under operation, the diesel generator has been operated close to its maximum capacity. The total power at *t* = *10,800 sec.* has reached its first peak point of 8.107 MVA as the generated power, and 7.435 MVA as the total load demand. The loads have met the minimum value of 5.616 MVA at *t* = *14,400 sec.* where the generated power has been 6.062 MVA. Between *t* = *14,400 sec.* and *t* = *16,550 sec.*, the total load demand has increased to 6.303 MVA (due to the different charging profiles), and the generation units have followed the load profile, and have generated 6.831 MVA. Over time, the total load level has shown an overall increasing trend and increased from 6.100 MVA at *t* = *18,800 sec.* to 6.317 MVA at *t* = *21,600 sec.* and 8.071 MVA at *t* = *28,800 sec.*, where the total generated power has reached 6.766 MVA, 6.802 MVA, 8.771 MVA, respectively. The power grid has experienced the second increase between *t* = *28,800 sec.* and *t* = *30.160 sec.* in which more cars have been in the charging mode. During this time interval, the total load demand has increased from 9.916 MVA to 10.200 MVA, and the generated power has been 10.680 MVA and 11.020 MVA, respectively. The same trend has been observed between *t* = *30,940 sec.* and *t* = *64,800 sec.* in which the number of PHEVs in charge has been increased. The demand has changed from 8.591 MVA to 10.270 MVA, and correspondingly, the generated power has varied from 9.365 MVA to 11.440 MVA, respectively. At the third peak point, the generated power has reached its maximum capacity where approximately the total load demand has been 12.650 MVA. From t = *66.200 sec.* to t = *79,200 sec.*, both the total load demand and generated power have decreased, but in general, their corresponding values have remained at a high level. At t = *79,200 sec.*, the power grid has lost 4.5 MW active power due to the outage of the wind farm and accordingly, there has been an intensive drop in the generated power, and the load demand has been more than the generated power. This could interrupt the power flow, and there has been this need to use an auxiliary system, such as the V2G system to restore power. As shown in Figure 20, PHEVs have supported the grid and contributed to the voltage and frequency regulation, and also the robustness of the proposed control system has been illustrated. The decreasing trend in both the generated power and total load demand has continued until the load level has decreased.

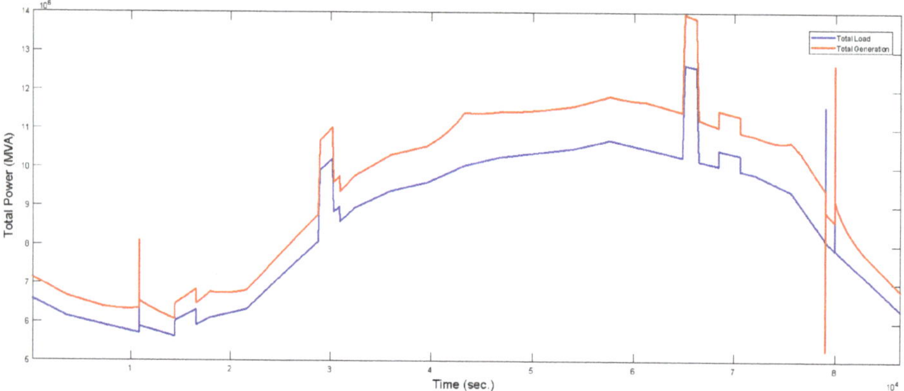

Figure 20. Total generated and consumed power curves in 24 h when the V2G system is under operation.

5. Conclusions

This paper reveals the impact of PHEVs charging on the power grid. A bidirectional charging station with a novel control strategy is proposed to solve the problem of voltage and frequency regulation in the power system due to the charging of PHEVs. A central AC/DC VSC converter station is investigated to inject active and reactive power into the power grid to regulate the voltage and frequency and reduce the peak load, as well as power quality improvement by considering the SoC and available active power in power grids. The proposed control strategy allows PHEVs to contribute to the grid regulation when an event occurs in a DG-based power grid consisting of different microgrids, diesel generator and wind farm, PHEVs with several charging profiles, and different loads. The simulation results show the robustness of the proposed control strategy.

Author Contributions: F.M. was responsible for methodology, collecting resources, data analysis, writing—original draft preparation, and writing—review and editing. G.-A.N. and M.S. were responsible for the supervision, and writing—review and editing.

Funding: M. Saif acknowledges the financial support of Natural Sciences and Engineering Research Council (NSERC) of Canada, as well as National Science Foundation of China under grant number 61873144.

Conflicts of Interest: The authors declare no conflicts of interest.

References

1. Mohammadi, F. Design and Electrification of an Electric Vehicle Using Lithium-ion Batteries. In Proceedings of the 3rd International Conference on Electrical Engineering, Tehran, Iran, 7 September 2018.
2. Mohammadi, F. Electric Vehicle Battery Market Analysis: Lead Acid. In Proceedings of the 9th Iranian Conference on Electrical and Electronics Engineering (ICEEE), Gonabad, Iran, 28 August 2018.
3. Mohammadi, F. Electric Vehicle Battery Market Analysis: Nickel Metal Hydride. In Proceedings of the 9th Iranian Conference on Electrical and Electronics Engineering (ICEEE), Gonabad, Iran, 28 August 2018.
4. Mohammadi, F. Electric Vehicle Battery Market Analysis: Lithium-ion. In Proceedings of the 1st International Conference on Modern Approaches in Engineering Science (ICMAES), Tbilisi, Georgia, 21 November 2018.
5. Mohammadi, F. Analysis and Electrification of the Solar-Powered Electric Vehicle. In Proceedings of the 5th Iranian Conference and Exhibition on Solar Energy (ICESE), Tehran, Iran, 18–21 August 2018.
6. Mohammadi, F. Research in the Past, Present and Future Solar Electric Aircraft. *J. Sol. Energy Res.* **2018**, *3*, 237–248.
7. Mohammadi, F. Design, Analysis, and Electrification of a Solar-Powered Electric Vehicle. *J. Sol. Energy Res.* **2018**, *3*, 293–299.
8. Mohammadi, F. Hybridization of an Electric Vehicle Using Lithium-ion Batteries. In Proceedings of the 1st International Conference on Modern Approaches in Engineering Science (ICMAES), Tbilisi, Georgia, 21 November 2018.
9. Bauer, P.; Zhou, Y.; Doppler, J.; Stembridge, N. Charging of Electric Vehicles and Impact on the Grid. In Proceedings of the 13th Mechatronika, Trencianske Teplice, Slovakia, 2–4 June 2010.
10. Dharmakeerthi, C.; Mithulananthan, H.N.; Saha, T.K. Modeling and Planning of EV Fast Charging Station in Power Grid. In Proceedings of the IEEE Power and Energy Society General Meeting, San Diego, CA, USA, 22–26 July 2012.
11. Ding, M.; Zhang, Y.; Mao, M. Key Technologies for Microgrids—A Review. In Proceedings of the International Conference on Sustainable Power Generation and Supply, Nanjing, China, 6–7 April 2009.
12. Pourbabak, H.; Kazemi, A. A New Technique for Islanding Detection Using Voltage Phase Angle of Inverter-Based DGs. *Int. J. Electr. Power Energy Syst.* **2014**, *57*, 198–205. [CrossRef]
13. Chen, D.; Xu, L. Autonomous DC Voltage Control of a DC Microgrid with Multiple Slack Terminals. *IEEE Trans. Power Syst.* **2012**, *27*, 1897–1905. [CrossRef]
14. Liu, X.; Wang, P.; Loh, P.C. A Hybrid AC/DC Micro-Grid. In Proceedings of the Conference International Power Engineering Conference (IPEC), Singapore, 27–29 October 2010.
15. Wang, P.; Liu, X.; Jin, C.; Loh, P.; Choo, F. A Hybrid AC/DC Micro-Grid Architecture, Operation and Control. In Proceedings of the IEEE Power and Energy Society General Meeting, San Diego, CA, USA, 24–29 July 2011.
16. Kakigano, H.; Miura, Y.; Ise, T. Low-Voltage Bipolar Type DC Microgrid for Super High Quality Distribution. *IEEE Trans. Power Electron.* **2010**, *25*, 3066–3075. [CrossRef]
17. Mohammadi, F. Power Management Strategy in Multi-Terminal VSC-HVDC System. In Proceedings of the 4th National Conference on Applied Research in Electrical, Mechanical Computer and IT Engineering, Tehran, Iran, 4 October 2018.
18. Mohammadi, F.; Zheng, C. Stability Analysis of Electric Power System. In Proceedings of the 4th National Conference on Technology in Electrical and Computer Engineering, Tehran, Iran, 27 December 2018.
19. Xu, L.; Chen, D. Control and Operation of a DC Microgrid with Variable Generation and Energy Storage. *IEEE Trans. Power Deliv.* **2011**, *26*, 2513–2522. [CrossRef]
20. Liu, X.; Wang, P.; Loh, P.C. A Hybrid AC/DC Microgrid and Its Coordination Control. *IEEE Trans. Smart Grid* **2011**, *2*, 278–286.
21. Nejabatkhah, F.; Li, Y.W. Overview of Power Management Strategies of Hybrid AC/DC Microgrid. *IEEE Trans. Power Electron.* **2014**, *30*, 7072–7089. [CrossRef]
22. Lu, X.; Guerrero, J.M.; Sun, K.; Vasquez, J.C. An Improved Droop Control Method for DC Microgrids Based on Low Bandwidth Communication with DC Bus Voltage Restoration and Enhanced Current Sharing Accuracy. *IEEE Trans. Power Electron.* **2014**, *29*, 1800–1812. [CrossRef]
23. Sahoo, S.; Zhang, C.; Pullaguram, D.R.; Mishra, S.; Wu, J.; Senroya, N. Investigation of Distributed Cooperative Control for DC Microgrids in Different Communication Medium. *Energy Procedia* **2017**, *142*, 2218–2223. [CrossRef]

24. Varghese, A.; Chandran, L.R.; Rajendran, A. Power Flow Control of Solar PV Based Islanded Low Voltage DC Microgrid with Battery Management System. In Proceedings of the IEEE 1st International Conference on Power Electronics, Intelligent Control and Energy Systems (ICPEICES), Delhi, India, 4–6 July 2016.
25. Guerrero, J.M.; Vasquez, J.C.; Matas, J.; de Vicuna, L.G.; Castilla, M. Hierarchical Control of Droop-Controlled AC and DC Microgrids-A General Approach Toward Standardization. *IEEE Trans. Ind. Electron.* **2010**, *58*, 158–172. [CrossRef]
26. Vasquez, J.C.; Guerrero, J.M.; Miret, J.; Castilla, M.; de Vicuna, L.G. Hierarchical Control of Intelligent Microgrids. *IEEE Ind. Electron. Mag.* **2010**, *4*, 23–29. [CrossRef]
27. Meng, L.; Savaghebi, M.; Andrade, F.; Vasquez, J.C.; Guerrero, J.M.; Graells, M. Microgrid Central Controller Development and Hierarchical Control Implementation in the Intelligent Microgrid Lab of Aalborg University. In Proceedings of the IEEE Applied Power Electronics Conference and Exposition (APEC), Charlotte, NC, USA, 15–19 March 2015.
28. Unamuno, E.; Barrena, J.A. Hybrid AC/DC Microgrids-Part II: Review and Classification of Control Strategies. *Renew. Sustain. Energy Rev.* **2015**, *52*, 1123–1134. [CrossRef]
29. Feng, X.; Shekhar, A.; Yang, F.; Hebner, R.E.; Bauer, P. Comparison of Hierarchical Control and Distributed Control for Microgrid. *Electr. Power Compon. Syst.* **2017**, *45*, 1043–1056. [CrossRef]
30. Marzal, S.R.; González-Medina, R.; Salas-Puente, R.; Figueres, E.; Garcerá, G. A Novel Locality Algorithm and Peer-to-Peer Communication Infrastructure for Optimizing Network Performance in Smart Microgrids. *Energies* **2017**, *10*, 1275.
31. Kaura, A.; Kaushalb, J.; Basakc, P. A Review on Microgrid Central Controller. *Renew. Sustain. Energy Rev.* **2016**, *55*, 338–345. [CrossRef]
32. Begum, M.; Abuhilaleh, M.; Li, L.; Zhu, J. Distributed Secondary Voltage Regulation for Autonomous Microgrid. In Proceedings of the 20th International Conference on Electrical Machines and Systems (ICEMS), Sydney, Australia, 11–14 August 2017.
33. Wu, D.; Tang, F.; Dragicevic, T.; Guerrero, J.M.; Vasquez, J.C. Coordinated Control Based on Bus-Signaling and Virtual Inertia for Islanded DC Microgrids. *IEEE Trans. Smart Grid* **2015**, *6*, 2627–2638. [CrossRef]
34. Shi, D.; Chen, X.; Wang, Z.; Zhang, X.; Yu, Z.; Wang, X.; Bian, D. A Distributed Cooperative Control Framework for Synchronized Reconnection of a Multi-Bus Microgrid. *IEEE Trans. Smart Grid* **2018**, *9*, 6646–6655. [CrossRef]
35. Dou, C.; Zhang, Z.; Yue, D.; Zheng, Y. MAS-Based Hierarchical Distributed Coordinate Control Strategy of Virtual Power Source Voltage in Low-Voltage Microgrid. *IEEE Access* **2017**, *5*, 11381–11390. [CrossRef]
36. Bidram, A.; Lewis, F.L.; Davoudi, A. Distributed Control Systems for Small-Scale Power Networks: Using Multiagent Cooperative Control Theory. *IEEE Control Syst. Mag.* **2014**, *34*, 56–77.
37. Fernandez, L.P.; San Roman, T.G.; Cossent, R.; Domingo, C.M.; Frias, P. Assessment of the Impact of Plug-in Electric Vehicles on Distribution Networks. *IEEE Trans. Power Syst.* **2011**, *26*, 206–213. [CrossRef]
38. Ying, H.H.; Han, H.J.; Jun, W.X.; Qi, T.W. Optimal Control Strategy of Vehicle-To-Grid for Modifying the Load Curve Based on Discrete Particle Swarm Algorithm. In Proceedings of the 4th International Conference on Electric Utility Deregulation and Restructuring and Power Technologies (DRPT), Weihai, China, 6–9 July 2011.
39. Jiang, B.; Fei, Y. Decentralized Scheduling of PEV On-Street Parking and Charging for Smart Grid Reactive Power Compensation. In Proceedings of the IEEE PES Innovative Smart Grid Technologies Conference (ISGT), Washington, DC, USA, 24–27 Feberuary 2013.
40. Falahi, M.; Chou, H.M.; Ehsani, M.; Xie, L.K.L. Butler-Purry, Potential Power Quality Benefits of Electric Vehicles. *IEEE Trans. Sustain. Energy* **2013**, *4*, 1016–1023.
41. Bao, K.; Li, S.; Zheng, H. Battery Charge and Discharge Control for Energy Management in EV and Utility Integration. In Proceedings of the IEEE Power and Energy Society General Meeting, San Diego, CA, USA, 22–26 July 2012.
42. Kisacikoglu, M.C.; Ozpineci, B.; Tolbert, L.M. Examination of a PHEV Bidirectional Charger System for V2G Reactive Power Compensation. In Proceedings of the Twenty-Fifth Annual IEEE Applied Power Electronics Conference and Exposition (APEC), Palm Springs, CA, USA, 21–25 Feberuary 2010.
43. Ferreira, R.J.; Miranda, L.M.; Araújo, R.E.; Lopes, J.P. A New Bi-Directional Charger for Vehicle-to-Grid Integration. In Proceedings of the 2nd IEEE PES International Conference and Exhibition on Innovative Smart Grid Technologies, Manchester, UK, 5–7 December 2011.

44. Kadurek, P.; Ioakimidis, C.; Ferrao, P. Electric Vehicles and Their Impact to the Electric Grid in Isolated Systems. In Proceedings of the International Conference on Power Engineering, Energy and Electrical Drives, Lisbon, Portugal, 18–20 March 2009.
45. Onar, O.; Khaligh, A. Grid Interactions and Stability Analysis of a Distributed Network with Plug-in Hybrid Electric Vehicle Loads. In Proceedings of the 25th Annual IEEE Applied Power Electronics Conference and Exposition (APEC), Palm Springs, CA, USA, 21–25 Feberuary 2010.
46. Erb, D.C.; Onar, O.; Khaligh, A. Bi-directional Charging Topologies for Plug-in Hybrid Electric Vehicles. In Proceedings of the 25th Annual IEEE Applied Power Electronics Conference and Exposition (APEC), Palm Springs, CA, USA, 21–25 Feberuary 2010.
47. Musavi, F.; Edington, M.; Eberle, W.; Dunford, W.G. Energy Efficiency in Plug-in Hybrid Electric Vehicle Chargers: Evaluation and Comparison of Front End AC-DC Topologies. In Proceedings of the IEEE Energy Conversion Congress and Exposition, Phoenix, AZ, USA, 17–22 September 2011.
48. Musavi, F.; Eberle, W.; Dunford, W.G. A Phase Shifted Semi-Bridgeless Boost Power Factor Corrected Converter for Plug in Hybrid Electric Vehicle Battery Chargers. In Proceedings of the 26th Annual IEEE Applied Power Electronics Conference and Exposition (APEC), Fort Worth, TX, USA, 6–11 March 2011.
49. Musavi, F.; Eberle, W.; Dunford, W.G. A High-Performance Single-Phase Bridgeless Interleaved PFC Converter for Plug-in Hybrid Electric Vehicle Battery Chargers. *IEEE Trans. Ind. Appl.* **2011**, *47*, 1833–1843. [CrossRef]
50. Sahu, B.; Rincon-Mora, G.A. A Low Voltage, Dynamic, Noninverting, Synchronous Buck-Boost Converter for Portable Applications. *IEEE Trans. Power Electron.* **2004**, *19*, 443–452. [CrossRef]
51. Huang, P.C.; Wu, W.Q.; Ho, H.H.; Chen, K.H. Hybrid Buck-Boost Feedforward and Reduced Average Inductor Current Techniques in Fast Line Transient and High-Efficiency Buck-Boost Converter. *IEEE Trans. Power Electron.* **2010**, *25*, 719–730. [CrossRef]
52. Waffler, S.; Kolar, J.W. A Novel Low-Loss Modulation Strategy for High-Power Bidirectional Buck + Boost Converters. *IEEE Trans. Power Electron.* **2009**, *24*, 1589–1599. [CrossRef]
53. Camara, M.B.; Gualous, H.; Gustin, F.; Berthon, A.; Dakyo, B. DC/DC Converter Design for Supercapacitor and Battery Power Management in Hybrid Vehicle Applications-Polynomial Control Strategy. *IEEE Trans. Ind. Electron.* **2010**, *57*, 587–597. [CrossRef]
54. Ghanem, M.C.; Al-Haddad, K.; Roy, G. A New Control Strategy to Achieve Sinusoidal Line Current in A Cascade Buck-Boost Converter. *IEEE Trans. Ind. Electron.* **1996**, *43*, 441–449. [CrossRef]
55. Gabriault, M.; Notman, A. A High Efficiency, Noninverting, Buck-Boost DC-DC Converter. In Proceedings of the 19th Annual IEEE Applied Power Electronics Conference and Exposition, Anaheim, CA, USA, 22–26 Feberuary 2004.
56. Midya, P.; Haddad, K.; Miller, M. Buck or Boost Tracking Power Controller. *IEEE Power Electron. Lett.* **2004**, *2*, 131–134. [CrossRef]
57. Hwu, K.I.; Yau, Y.T. Two Types of KY Buck-Boost Converters. *IEEE Trans. Ind. Electron.* **2009**, *56*, 2970–2980. [CrossRef]
58. Xu, D.; Zhao, C.; Fan, H. A PWM Plus Phase-Shift Control Bidirectional DC-DC Converter. *IEEE Trans. Power Electron.* **2004**, *19*, 666–675. [CrossRef]
59. Gang, M.; Yuanyuan, L.; Wenlong, Q. A Novel Soft Switching Bidirectional DC/DC Converter and Its Output Characteristic. In Proceedings of the IEEE Region 10 Conference, Hong Kong, China, 14–17 November 2006.
60. Krismer, F.; Biela, J.; Kolar, J.W. A Comparative Evaluation of Isolated Bi-directional DC/DC Converters with Wide Input and Output Voltage Range. In Proceedings of the 14th IAS Annual Meeting on Conference Record of the 2005 Industry Applications Conference, Hong Kong, China, 2–6 October 2005; IEEE: Piscataway, NJ, USA, 2005.
61. Tsuruta, Y.; Ito, Y.; Kawamura, A. Snubber-Assisted Zero-Voltage and Zero-Current Transition Bilateral Buck and Boost Chopper for EV Drive Application and Test Evaluation at 25 kW. *IEEE Trans. Ind. Electron.* **2009**, *56*, 4–11. [CrossRef]
62. Jung, D.Y.; Hwang, S.H.; Ji, Y.H.; Lee, J.H.; Jung, Y.C.; Won, C.Y. Soft-Switching Bidirectional DC/DC Converter with a LC Series Resonant Circuit. *IEEE Trans. Power Electron.* **2013**, *28*, 1680–1690. [CrossRef]
63. Wu, H.; Lu, J.; Shi, W.; Xing, Y. Nonisolated Bidirectional DC-DC Converters with Negative-Coupled Inductor. *IEEE Trans. Power Electron.* **2012**, *27*, 2231–2235. [CrossRef]

64. Do, H.L. Nonisolated Bidirectional Zero-Voltage-Switching DC-DC Converter. *IEEE Trans. Power Electron.* **2011**, *26*, 2563–2569. [CrossRef]
65. Arancibia, A.; Strunz, K. Modeling of an Electric Vehicle Charging Station for Fast DC Charging. In Proceedings of the IEEE International Electric Vehicle Conference, Greenville, SC, USA, 4–8 March 2012.
66. Aggeler, D.; Canales, F.; Parra, H.Z.D.L.; Coccia, A.; Butcher, N.; Apeldoorn, O. Ultra-Fast DC-Charge Infrastructures for EV-Mobility and Future Smart Grids. In Proceedings of the IEEE PES Innovative Smart Grid Technologies Conference Europe (ISGT Europe), Gothenberg, Sweden, 11–13 October 2010.
67. Gamboa, G.; Hamilton, C.; Kerley, R.; Elmes, S.; Arias, A.; Shen, J.; Batarseh, I. Control Strategy of a Multi-Port, Grid Connected, Direct-DC PV Charging Station for Plug-in Electric Vehicles. In Proceedings of the IEEE Energy Conversion Congress and Exposition, Phoenix, AZ, USA, 17–22 September 2010.
68. Takagi, M.; Yamaji, K.; Yamamoto, H. Power System Stabilization by Charging Power Management of Plug-in Hybrid Electric Vehicles with LFC Signal. In Proceedings of the IEEE Vehicle Power and Propulsion Conference, Dearborn, MI, USA, 7–10 September 2009.
69. Ota, Y.; Taniguchi, H.; Nakajima, T.K.; Liyanage, M.; Baba, J.; Yokoyama, A. Autonomous Distributed V2G (Vehicle-to-Grid) Considering Charging Request and Battery Condition. In Proceedings of the IEEE PES Innovative Smart Grid Technologies Conference Europe (ISGT Europe), Gothenberg, Sweden, 11–13 October 2010.
70. Ota, Y.; Taniguchi, H.; Nakajima, T.; Liyanage, K.M.; Baba, J.; Yokoyama, A. Autonomous Distributed V2G (Vehicle-to-Grid) Satisfying Scheduled Charging. *IEEE Trans. Smart Grid* **2012**, *3*, 559–564. [CrossRef]
71. Li, C.T.; Ahn, C.; Peng, H.; Sun, J. Synergistic Control of Plug-in Vehicle Charging and Wind Power Scheduling. *IEEE Trans. Power Syst.* **2013**, *28*, 1113–1121. [CrossRef]
72. Yang, H.; Chung, C.Y.; Zhao, J. Application of Plug-in Electric Vehicles to Frequency Regulation Based on Distributed Signal Acquisition via Limited Communication. *IEEE Trans. Power Syst.* **2013**, *28*, 1017–1026. [CrossRef]
73. Liu, H.; Hu, Z.; Song, Y.; Lin, J. Decentralized Vehicle-to-Grid Control for Primary Frequency Regulation Considering Charging Demands. *IEEE Trans. Power Syst.* **2013**, *28*, 3480–3489. [CrossRef]
74. Masuta, T.; Yokoyama, A. Supplementary Load Frequency Control by Use of a Number of Both Electric Vehicles and Heat Pump Water Heaters. *IEEE Trans. Smart Grid* **2012**, *3*, 1253–1262. [CrossRef]
75. Arya, Y.; Kumar, N.; Mathur, H.D. Automatic Generation Control in Multi Area Interconnected Power System by using HVDC Links. *Int. J. Power Electron. Drive Syst.* **2012**, *2*, 67. [CrossRef]
76. Vachirasricirikul, S.; Ngamroo, I. Robust LFC in a Smart Grid with Wind Power Penetration by Coordinated V2G Control and Frequency Controller. *IEEE Trans. Smart Grid* **2014**, *5*, 371–380. [CrossRef]
77. Jayakumar, A.; Chalmers, A.; Lie, T.T. Review of Prospects for Adoption of Fuel Cell Electric Vehicles in New Zealand. *IET Electr. Syst. Transp.* **2017**, *7*, 259–266. [CrossRef]
78. Mathworks. 24-Hour Simulation of a Vehicle-to-Grid (V2G) System. Available online: https://www.mathworks.com/help/physmod/sps/examples/24-hour-simulation-of-a-vehicle-to-grid-v2g-system.html (accessed on 1 April 2019).

Article

A New Topology of a Fast Proactive Hybrid DC Circuit Breaker for MT-HVDC Grids

Fazel Mohammadi [1,*], Gholam-Abbas Nazri [2] and Mehrdad Saif [1]

1 Electrical and Computer Engineering (ECE) Department, University of Windsor, Windsor, ON N9B 1K3, Canada
2 Electrical and Computer Engineering, Wayne State University, Detroit, MI 48202, USA
* Correspondence: fazel@uwindsor.ca or fazel.mohammadi@ieee.org

Received: 26 June 2019; Accepted: 14 August 2019; Published: 19 August 2019

Abstract: One of the major challenges toward the reliable and safe operation of the Multi-Terminal HVDC (MT-HVDC) grids arises from the need for a very fast DC-side protection system to detect, identify, and interrupt the DC faults. Utilizing DC Circuit Breakers (CBs) to isolate the faulty line and using a converter topology to interrupt the DC fault current are the two practical ways to clear the DC fault without causing a large loss of power infeed. This paper presents a new topology of a fast proactive Hybrid DC Circuit Breaker (HDCCB) to isolate the DC faults in MT-HVDC grids in case of fault current interruption, along with lowering the conduction losses and lowering the interruption time. The proposed topology is based on the inverse current injection technique using a diode and a capacitor to enforce the fault current to zero. Also, in case of bidirectional fault current interruption, the diode and capacitor prevent changing their polarities after identifying the direction of fault current, and this can be used to reduce the interruption time accordingly. Different modes of operation of the proposed topology are presented in detail and tested in a simulation-based system. Compared to the conventional DC CB, the proposed topology has increased the breaking current capability, and reduced the interruption time, as well as lowering the on-state switching power losses. To check and verify the performance and efficiency of the proposed topology, a DC-link representing a DC-pole of an MT-HVDC system is simulated and analyzed in the PSCAD/EMTDC environment. The simulation results verify the robustness and effectiveness of the proposed HDCCB in improving the overall performance of MT-HVDC systems and increasing the reliability of the DC grids.

Keywords: DC Circuit Breaker (CB); DC Fault; Hybrid DC Circuit Breaker (HDCCB); Multi-Terminal VSC-HVDC (MT-HVDC) Grids

1. Introduction

Protection of the MT HVDC grids in case of a DC fault is vital to improve the reliability of the power system [1–3]. Conventional Voltage Sourced Converters (VSCs) are not capable of limiting or interrupting the current that flows following a DC-side fault. Also, the high power DC Circuit Breakers (CBs) are not commercially available. Therefore, opening the CB on the AC side is the only possible way to isolate the DC-side fault. Opening all the AC CBs to isolate/clear the DC-side fault in a DC grid with high capacity can lead to a large loss of power infeed. Therefore, clearing the DC fault within the DC grid by opening the AC CBs is practically impossible [4–6].

The consequence of a DC-side fault within an MT-HVDC grid is the sudden increase in the fault current, and that is due to the capacitive behavior of the HVDC grids and low resistance of the DC cables. The fault current should be interrupted in less than 20 ms to limit it within the acceptable levels [7]. Using AC-side CB in the point-to-point VSC-HVDC grids to interrupt the fault current leads to loss of the entire link. Adopting the same strategy for MT-HVDC grids and opening all the AC-side CBs is a disruptive measure and can cause losing a large amount of power infeed. Therefore, it is mandatory to isolate only the faulty section of MT-HVDC grids. Two measures can be considered to deal with this issue: (1) Using fast-acting DC CBs at both ends of the DC-links. (2) Using fault-blocking converters to open the DC-links in case of a sudden increase in the DC fault current. Moreover, determining the exact location of the DC-side faults within MT-HVDC grids is difficult [8,9]. In the AC grids, impedance relays are used to determine the fault location. However, due to the low resistance of the DC cables compared to the impedance of the AC grids, it is impractical for the impedance relay to determine the fault location in a DC-link within its protective zone. Besides, distinguishing between the AC and DC-side faults is another major challenge in a DC protection system [10,11].

In order to interrupt the fault current, the AC CBs at the high voltage transmission levels require 4-5 cycles (within 80–100 ms). By creating zero crossings of the current, the AC CBs provide interruption with minimal arcing [12]. As there is no zero crossing of the current in DC, the conventional AC CBs are not useful to prevent large DC currents. To overcome this issue, a passive or active resonant circuit can be used to force the current zero crossing. The other way is to bypass or interrupt the fault current using semiconductor switches (Insulated-Gate Bipolar Transistor (IGBTs)), which considering the power losses, it is an expensive strategy. Due to the lack of reactance in the DC circuits, the rate of rise of fault current in the DC grids is higher than the AC grids [13,14]. Thus, the DC CBs must be capable of interrupting the fault currents faster than the AC CBs.

Different CB topologies are studied in the literature. The CBs can be categorized into three types: Mechanical Circuit Breakers (MCBs), Solid-State Circuit Breakers (SSCBs), and Hybrid Circuit Breakers (HCBs) [15,16]. The MCBs consist of the conventional AC CBs with a parallel circuit to generate a current zero crossing [17]. In both passive and active MCBs structures, due to the existence of the parallel circuit, the interruption time of the MCBs is long (30–50 ms), which is not suitable for VSC-HVDC systems. The SSCBs consist of several solid-state switches that can interrupt the fault current faster than the MCBs without the need for a current zero crossing [18–21]. Proper configuration of switches can lead to achieving the desired breaking current capability. To design a bidirectional CB, at least two switches are needed. The SSCBs are faster compared to the MCBs, but their on-state switching losses and total cost are high. The HCBs configurations are the results of the combination of MCBs and SSCBs [22]. The combinations are based on the operation time, breaking current capability, power losses, and total cost. The operation time and breaking current capability of the MCBs are low, but they are cheap. In contrast, the SSCBs are fast, but when it comes to a complicated configuration for the VSC-HVDC systems, they are expensive. Also, because of the existence of permanent resistance, the SSCBs have large power losses compared to the MCBs. The HCBs take advantage of both MCBs and SSCBs, such as fast current interruption and low power losses [23,24].

The first HCB [24], which was based on inverse voltage generation method, provided the advantages of both MCBs and SSCBs. It consisted of a semiconductor-based CB using IGBTs and a bypass circuit developed by the semiconductor-based load communication switches in series with an ultra-fast disconnector. Reducing the fault current level to zero is achieved by generating the arc voltage at a higher level than the source voltage. This topology has a short interruption time than the MCSs and lower on-state switching losses compared to the SSCBs. A detailed analysis of the operating modes of HCBs (reclosing and rebreaking) is performed in [16]. An accurate model considering the coordinated control of the four subunits (two semiconductor values along with two mechanical switches) and the opening and closing sequences of each subunit of the HCB is presented in [23]. This topology is not suitable for MT-HVDC grids, where the fault location is hard to find, and the switches in the auxiliary CB might damage due to the high rate of rise of current. To overcome this issue, a large number of semiconductor switches in the semiconductor-based CB branch is needed, which can increase the cost of implementation and relatively increase the on-state switching losses. To reduce the on-state switching losses of the HCBs, another topology considering multiple thyristors connected in series is proposed in [25]. In this topology, by injecting current stored in the capacitor with the initial charge, a current zero crossing can be generated. The main issue with this topology is the high on-state switching losses because of the existence of IGBT in the main CB. Moreover, the fault interruption time is increased due to the transferring the fault current from the CB to the auxiliary CB. Besides, using an external power supply to charge the capacitor might lead to an increase in the costs of implementation. This topology provides the bidirectional fault current interruption, but reclosing and rebreaking capabilities of the HCBs is not considered. The topology of the HCBs presented in [26] is provided reclosing and rebreaking capabilities, as well as bidirectional fault current interruption. Reclosing and rebreaking capabilities have prevented the interruption of the power to the connected AC systems, and the auxiliary power source is used to charge the capacitor. Although the polarity of the capacitor can change according to the direction of fault current, it might have delay time to interrupt the fault and consequently, increase of the fault current magnitude.

As MT-HVDC systems can control the active and reactive power independently and in both directions [27], and do not require the installation of separate Flexible AC Transmission System (FACTS) devices, it is necessary to investigate a proper Hybrid DC Circuit Breaker (HDCCB) for MT-HVDC systems with capabilities of reclosing and rebreaking, as well as bidirectional fault current interruption. A new topology of the HDCCB to solve the above-mentioned issues is proposed.

The major contributions of this paper are as follows:

- The proposed topology of the HDCCB is based on the switching technique to minimize the costs of implementation and on-state switching losses, as well as interrupting the fault current in both directions. Also, in this topology, the polarity of the capacitor is based on the direction of fault current, which would lead to reducing the interruption time. Hence, it can be used in MT-HVDC systems.
- The proposed topology has reclosing and rebreaking capabilities without the need for an external power supply, which leads to reducing the overall cost.
- Improving the fault tolerance capability by increasing the maximum breaking current 0.25% compared to the conventional HDCCB, while having approximately the same total dissipated energy of the surge arrestor in the DC CB, and clearing the fault in 16 ms.
- The proposed topology limits the rate of rise of the voltage across the DC CB, reduces the on-state switching losses, and ensures an equal voltage distribution regardless of tolerances in the switching characteristics.
- The proposed topology can improve the overall performance of MT-HVDC systems and increase the reliability of the DC grids.

2. Configuration and Operation of the Proposed Hybrid DC Circuit Breaker

2.1. Configuration of the Proposed Hybrid DC Circuit Breaker

Figure 1 illustrates the overall structure of the proposed HDCCB with bidirectional current interruption capability. This structure consists of two main DC CBs in the main branch, a fast mechanical switch and an auxiliary DC CB in the auxiliary branch, and a residual DC current disconnector. The residual DC current disconnector (in series with a current limiting reactor) is used to entirely isolate the DC circuit.

Figure 2 demonstrates the proposed configuration of the main DC CBs in the main branch. In this configuration, four submodules are considered to be connected back-to-back to isolate the DC current fault. Each submodule includes a diode, D, and a low resistance switch, S, in series and a capacitor, C, in parallel with the diode and switch. The switches are capable of breaking the bidirectional fault current. Also, the polarity of the capacitor changes with respect to the direction of the fault current. In order to protect the back-to-back submodules and prevent the capacitors from overvoltage, a Metal Oxide Varistor, MOV, is used.

The auxiliary DC CB consists of two switches, as shown in Figure 3. This CB matches lower voltage and current capability.

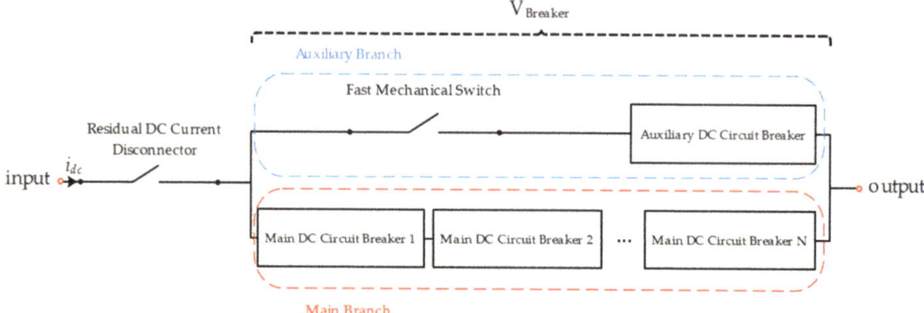

Figure 1. Overall structure of the proposed HDCCB.

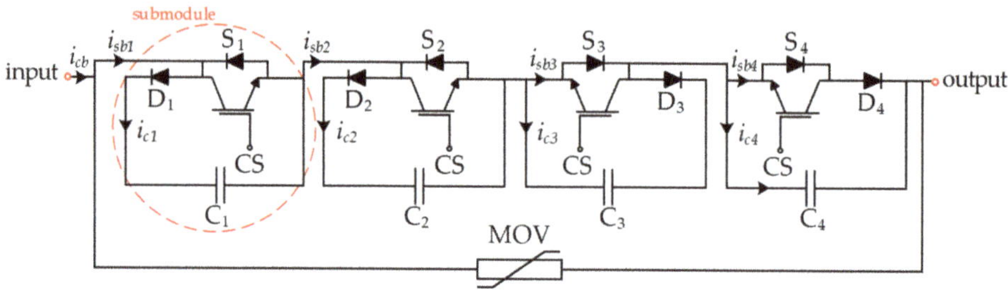

Figure 2. Configuration of the main DC CBs in the main branch.

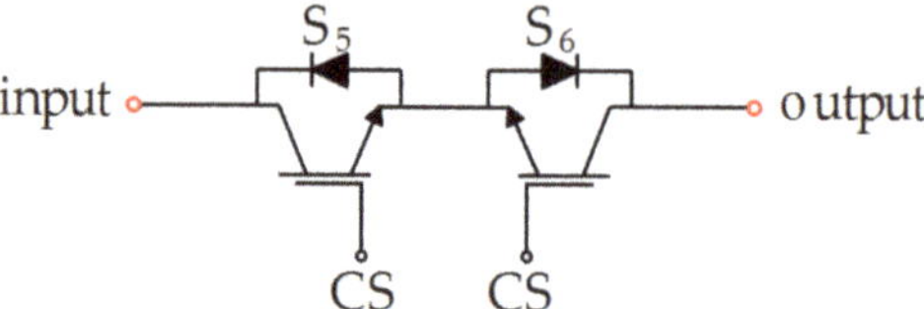

Figure 3. Configuration of the auxiliary DC CB.

2.2. Principle Operation of the Proposed Hybrid DC Circuit Breaker

Predefined states of each component in the submodule determine the operation modes of the proposed HDCCB. Considering the operation modes, the switches can be either turned on or off. During the normal operation, the DC current flows through the auxiliary branch. In case of the DC fault, the auxiliary DC CB commutes the current to the main branch, and the fast mechanical switch opens. During the current breaking, the fast mechanical switch isolates the auxiliary DC CB from the voltage across the main DC CB. As a result, the rating voltage of the auxiliary DC CB reduces. Therefore, the majority of the power losses would be related to the switching losses. The fast mechanical switch opens when zero current and/or low voltage stress can be detected. The recovery voltage of the fast mechanical switch determines by the proactive level of the surge arresters, while both the main and auxiliary DC CBs are open. By connecting several mechanical switches, the operation time exceeds 30 ms. Proper control of the HDCCB leads to minimizing the time delay of the mechanical switching elements. Hence, the operation time of the multiple switches should be less than the time for selective protection. Figure 4 depicts the typical proactive control of the HDCCB. The proactive control system enables when the level of the DC line current exceeds the predefined overcurrent level. In this case, two events are expected: (1) The selective protection sends a trip signal. (2) The DC current of the faulty line reaches the maximum breaking current capability of the main DC CB. In both cases, there would be a time delay for the current breaking of the main DC CB.

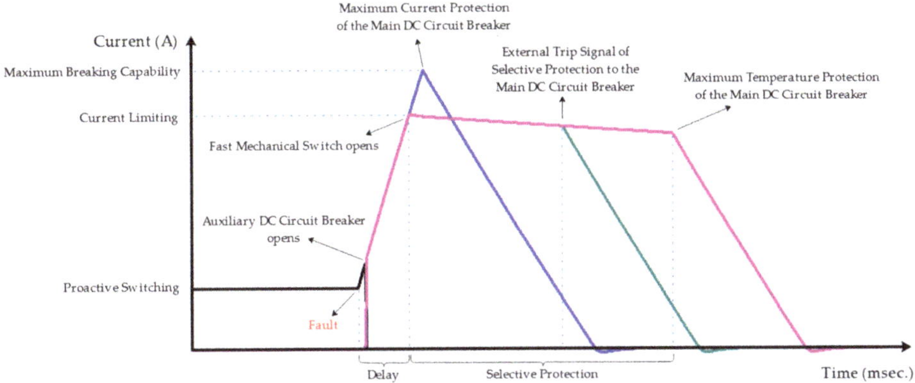

Figure 4. The current waveform of a typical proactive HDCCB.

To extend the operation time until the main DC CB sends the trip signal, the main DC CB may disconnect the line in the current limitation mode prior to the current breaking. To prevent sudden increases in the line current, the voltage drop across the DC reactor should be controlled by the main DC CB. The energy dissipation capability of the surge arrester determines the maximum operation time of the current limiting mode. Also, to allow maintenance on demand while not interrupting the power transfer in the DC grids, scheduled current transfer of the line current from the auxiliary DC CB to the main DC CB is required. Considering the proactive mode, overcurrent in the DC grid can activate the current transfer from the auxiliary branch into the main branch of the DC CB prior to the trip signal of the protection device. Considering the CB failure, the backup CB activates within less than 0.2 ms to avoid large disturbances in the DC grids and maintain the breaking current capability of the backup CB at the reasonable levels. After clearing the fault, the HDCCB returns to its normal operation mode.

According to Figure 2, when the DC CB, which is located at the power receiving, is closed, the power flows from the output of the DC CB, based on the unidirectional conduction ability of the diode that is paralleled with the IGBT in the submodule. When the DC CB, which is placed at the power sending end, is closed, the power flows from the input of the DC CB with the conduction of the IGBT in the submodule. Under the normal operation, the IGBT in the submodule is conducted and cascaded into the circuit. In case of DC fault, the IGBT of the submodule is shut down to prevent the diode and capacitor from being cascaded into the fault loop. By charging the capacitor, the back Electromotive Force (EMF) forces both ends of the equivalent capacitor on the short-circuit path to surpass the peak line voltage on the AC side to block the fault current. The fault tolerance capability can be significantly improved as the diode in the submodule, and the arm inductance in the AC/DC converter station in the DC fault are both shock resistance. This structure limits the rate of rise of the voltage across the DC CB, reduces the on-state switching losses, and guarantees an equal voltage distribution regardless of tolerances in the switching characteristics. Table 1 shows the operation status of the components in each submodule.

Table 1. The operation status of the components in each submodule.

Operation Mode	Diode	IGBT	Capacitor
Mode 1: Power Receiving End	Bypass	No Action	Bypass
Mode 2: Power Sending End	Bypass	Conducted	Bypass
Mode 3: Shut Down	Accessing to the Circuit	Shut Down	Charging

3. Design Parameters in the Proposed Hybrid DC Circuit Breaker

The designed parameters of the proposed HDCCB should be properly determined successfully to interrupt the fault current. Figure 5 illustrates the equivalent circuit of the DC CB in the DC grid in case of a short-circuit fault. As shown in this figure, the AC/DC converter station can be modeled as a DC voltage source, U_{DC}, and the line reactor in the DC grid is replaced by L_{DC} [28].

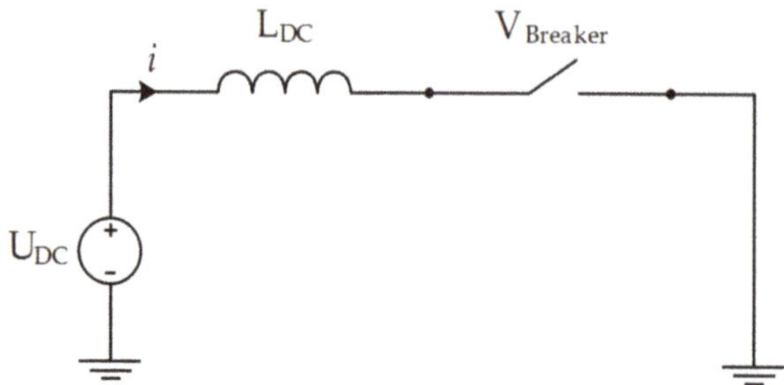

Figure 5. Equivalent circuit of the DC CB in the DC grid.

After the DC fault occurrence, the induced voltage, V_L, is generated by the inductor of the DC grid. Considering that the inter-terminal voltage of the DC CB is $V_{Breaker}$, then:

$$U_{DC} = V_L + V_{Breaker} = L_{DC}\frac{di}{dt} \tag{1}$$

By establishing the transient state of the voltage of the DC CB, V_L reverses, and $V_{Breaker}$ limits by the voltage of MOV, V_{MOV}. Therefore:

$$U_{DC} = V_L + V_{Breaker} = L_{DC}\frac{di}{dt} + V_{MOV} \tag{2}$$

According to Equations (1) and (2), the fault current can be derived as follows:

$$i = \begin{cases} \frac{U_{DC}}{L_{DC}}.t \\ \frac{U_{DC}-V_{MOV}}{L_{DC}}.t \end{cases} \tag{3}$$

Hence, the rate of rise of fault current can be calculated as follows:

$$\frac{di}{dt} = \begin{cases} \frac{U_{DC}}{L_{DC}} \\ \frac{U_{DC}-V_{MOV}}{L_{DC}} \end{cases} \tag{4}$$

Considering the time of the fault, t_{fault}, the maximum breaking current capability of the DC CB is as follows:

$$i_{(max)} = \frac{di}{dt}\cdot t_{fault} = \frac{U_{DC}}{L_{DC}}\cdot t_{fault} \tag{5}$$

In addition, the maximum rate of rise of fault current can be derived as follows:

$$\frac{di}{dt}_{(max)} = \frac{U_{DC}}{L_{DC}} \tag{6}$$

The energy dissipation capability of the HDCCB is from the DC power source and the stored energy in the line reactor. The energy dissipation from the DC power supply can be written as follows:

$$E_{source} = \int U_{DC}\cdot i dt = \int U_{DC}\cdot\frac{U_{DC} - V_{MOV}}{L_{DC}}\cdot t dt \tag{7}$$

In addition, the energy dissipation from the line reactor can be written as follows:

$$E_{reactor} = \frac{1}{2} L_{DC} (i_{(max)})^2 \tag{8}$$

The summation of the energy dissipation from the DC power source and the line reactor gives the total energy dissipation of the HDCCB.

$$E_{MOV} = \frac{1}{2} \cdot \frac{U_{DC} V_{MOV}}{U_{MOV} - V_{DC}} \cdot \left(\frac{1}{\frac{di}{dt}_{(max)}} \cdot i_{(max)} \right)^2 \tag{9}$$

4. Results and Discussions

4.1. Simulation Results

In order to check the performance of the proposed HDCCB, a DC-link representing a DC-pole of the VSC-HVDC system is modeled on PSCAD/EMTDC environment, as shown in Figure 6. An HDCCB is located at the end of the DC-link. As one of the most common faults in the VSC-HVDC systems, a single line-to-ground fault on the DC-link is selected for the fault mechanism analysis. It is assumed that the output voltage of the VSC-HVDC converter station, U_{DC}, is constant. Tables 2 and 3 show the main parameters of the DC-link and the specifications of the proposed HDCCB, respectively.

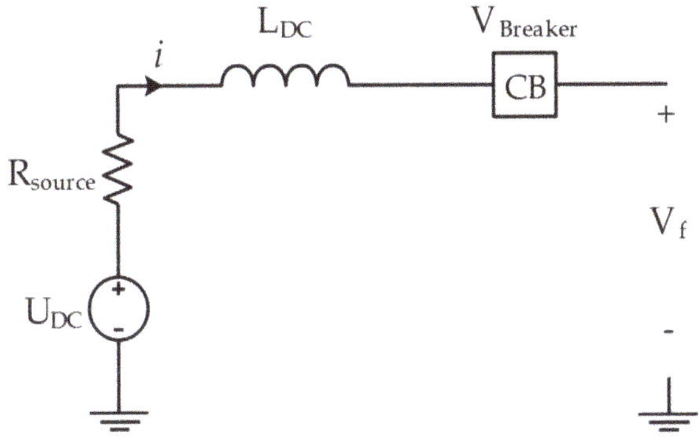

Figure 6. Configuration of the DC-link for fault analysis.

Table 2. DC-link parameters.

Parameter	Value
U_{DC}	320 kV
R_{source}	0.001 Ω
L_{DC}	100 mH

Table 3. Specifications of the proposed HDCCB.

Parameter	Value
Capacitor (C)	6.8 μF
Time delay of residual disconnector	20 ms
Time delay of fast mechanical switch	2 ms
Forward Breakover Voltage (IGBT)	100 kV
Forward Breakover Voltage (Diode)	100 kV

4.1.1. Rated Current Interruption

To give a general perspective about the relation of the designed parameters, considering 2 ms as the breaking time, and the fact that DC line fault is close to the DC switchyard, the maximum rate of rise of fault current is 3.5 kA/ms for a DC reactor of 100 mH in a 320 kV DC power grid with 10% maximum overvoltage. If the rated line current is 2 kA, the minimum required current breaking current capability of the HDCCB should be 9 kA. When there is no fault in the DC grid, the voltage drop across the DC CB is 0.04 V, and the total current passing through the DC grid is 1.6 kA. Figure 7 shows the behavior of the DC line current, and the current passing through the capacitors and switches of the HDCCB after applying a short-circuit DC fault at 5 s. As shown in Figure 7, the breaking time is less than 2.5 ms. The maximum current passing through the capacitors and switches of the HDCCB is 2.3157 kA and 8.0713 kA, respectively. The auxiliary CB opens into 50 µs. At 5.00195 s., the fast mechanical switch opens, where the corresponding value of the current limiting is 8.0713 kA. According to Figure 7, the maximum value of the DC fault current is 8.4425 kV, which is less than the minimum required breaking current capability of the HDCCB and the fault is successfully cleared within 16 ms. Based on the provided results, the maximum breaking current capability of 9 kA is confirmed. It should be noted that in Figure 7, the DC current passing through the line and the DC current of the proposed HDCCB are the same.

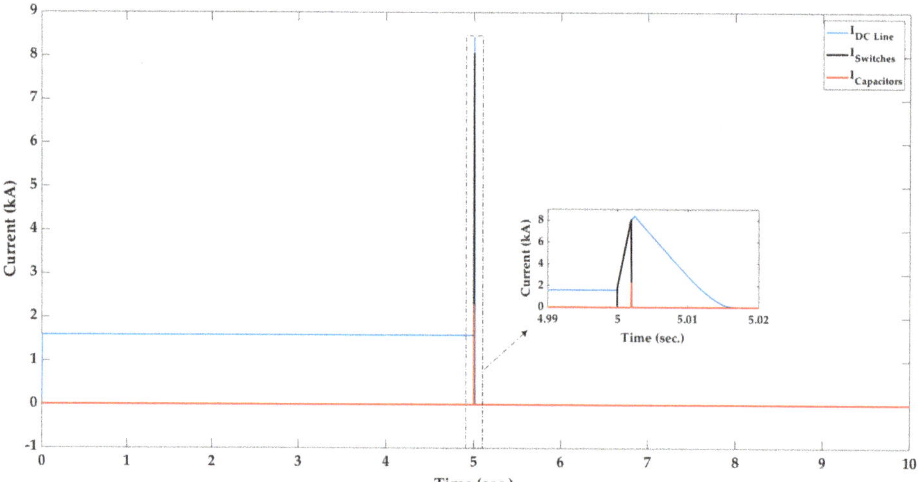

Figure 7. The DC line current, and the current passing through the capacitors and switches of the proposed HDCCB after applying a short-circuit DC fault at 5 s.

As there are some limitations in terms of breaking current capabilities of the IGBTs, by rising the current, voltage drop across the IGBTs increases significantly. To overcome this issue, the press-back IGBTs are used to ensure a reliable short-circuit without mechanical damage and failure of the IGBTs' modules. Therefore, in case of the failure of one of the IGBT modules, the fault can be cleared by the rest of IGBTs.

Figure 8 illustrates the CB voltage and voltage at the fault location (V_{Fault}) after applying a short-circuit DC fault at 5 s. As shown in Figure 8, the voltage across the proposed HDCCB exceeds approximately 180 kV during the current commutation, and the proposed HDCCB successfully clears the fault within 16 ms.

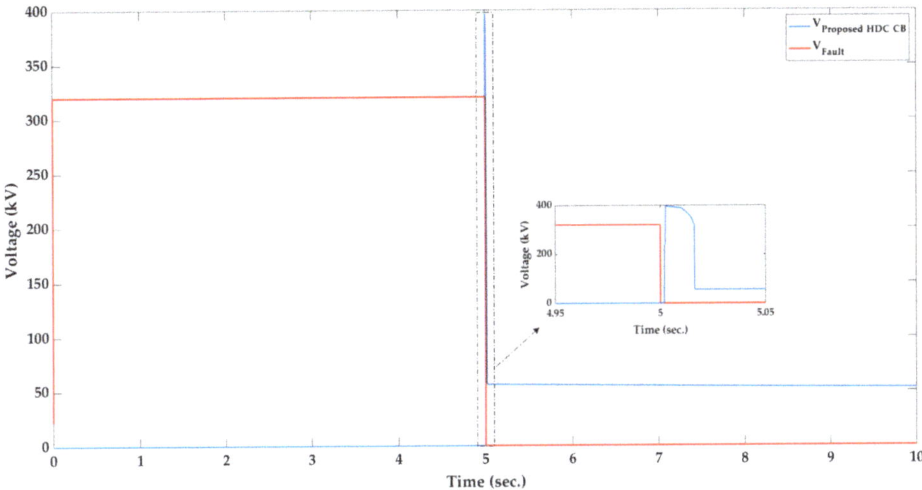

Figure 8. The CB voltage and voltage at the fault location after applying a short-circuit DC fault at 5 s.

By increasing the capacity of the capacitors of the main DC CBs, the maximum breaking current capability of the HDCCB increases. However, due to the discharge time of the capacitors, it takes longer time to entirely clear the DC fault.

Figure 9 demonstrates the dissipated energy of the surge arrestors in the two main DC CBs of the proposed HDCCB. The maximum dissipated energy of the surge arrestors in the main DC CB 1 and 2 is 12.6209 kJ and 7.7952 kJ, respectively.

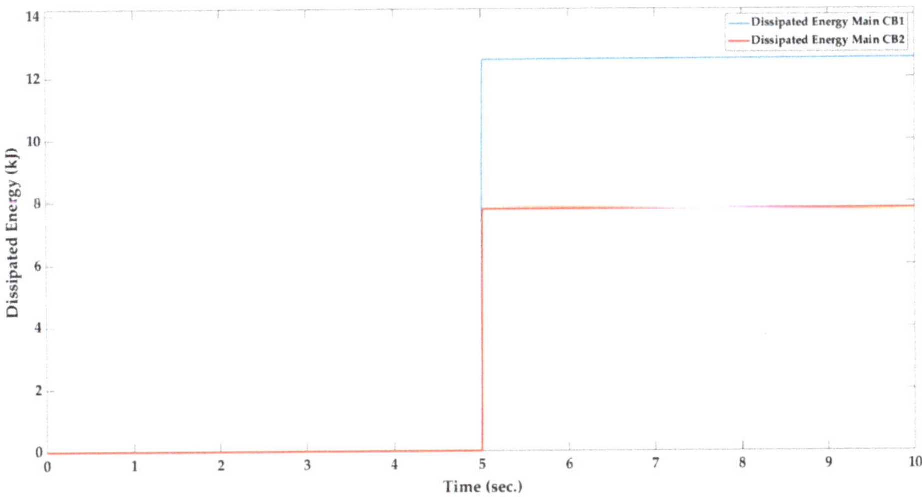

Figure 9. The dissipated energy of the surge arrestors in the two main DC CBs of the proposed HDCCB after applying a short-circuit DC fault at 5 s.

4.1.2. Reverse Current Interruption

In order to demonstrate the reverse current interruption of the proposed HDCCB, the direction of designed CB is reversed, so that the fault current flows in the opposite direction. Figure 10 shows that the same as the rated current interruption, the maximum breaking current capability of −9 kA is confirmed.

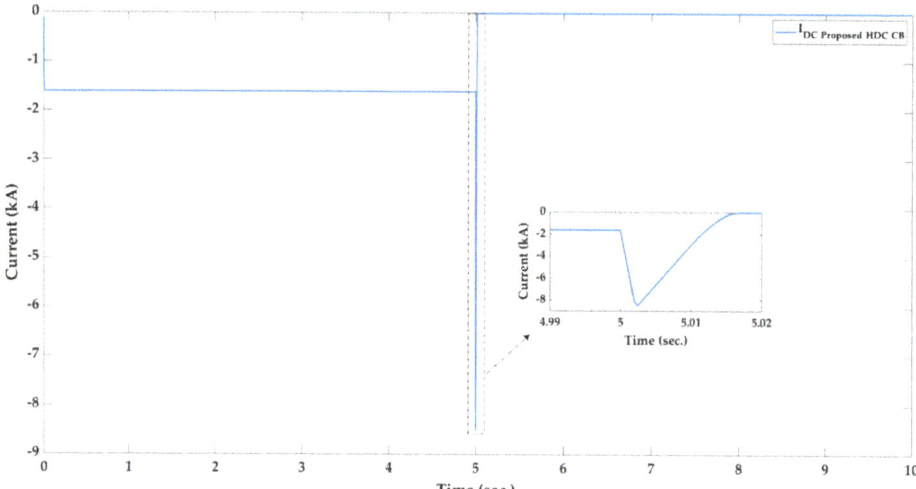

Figure 10. The DC current of the proposed HDCCB after applying a short-circuit DC fault at 5 s after changing the direction of the proposed HDCCB.

Figure 11 illustrates the CB voltage and voltage at the fault location after applying a short-circuit DC fault at 5 s after changing the direction of the proposed HDCCB. As shown in Figure 11, the proposed HDCCB successfully clears the fault within 16 ms.

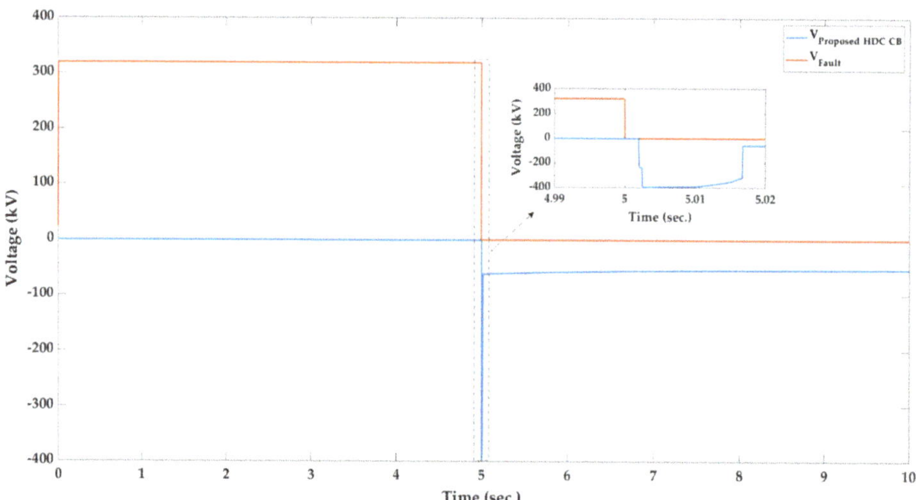

Figure 11. The CB voltage and voltage at the fault location after applying a short-circuit DC fault at 5 s after changing the direction of the proposed HDCCB.

4.1.3. Reclosing and Rebreaking Capabilities

To evaluate reclosing and rebreaking capabilities of the proposed HDCCB, a short time (dead time) between the first trip signal and the closing signal is considered. During this period of the time, there is no response from the proposed HDCCB to the test system. After the first trip signal, the proposed HDCCB opens and performs the first current interruption. The reclosing signal is sent to the proposed HDCCB 50 ms after the first trip, and the proposed HDCCB attempts to reconnect the source to the line. As the fault is not cleared from the system, the fault current rises as soon as the CB is closed. The second trip signal is sent to the proposed HDCCB 50 ms after the CB is closed, and then the proposed HDCCB performs the second current interruption. After that, the proposed HDCCB remains open to insulate the DC fault. Figures 12–15 show reclosing and rebreaking capabilities of the proposed HDCCB in both the rated and reverse current interruptions. The simulation results validate the reclosing and rebreaking capabilities of the proposed HDCCB in both the rated and reverse current interruptions.

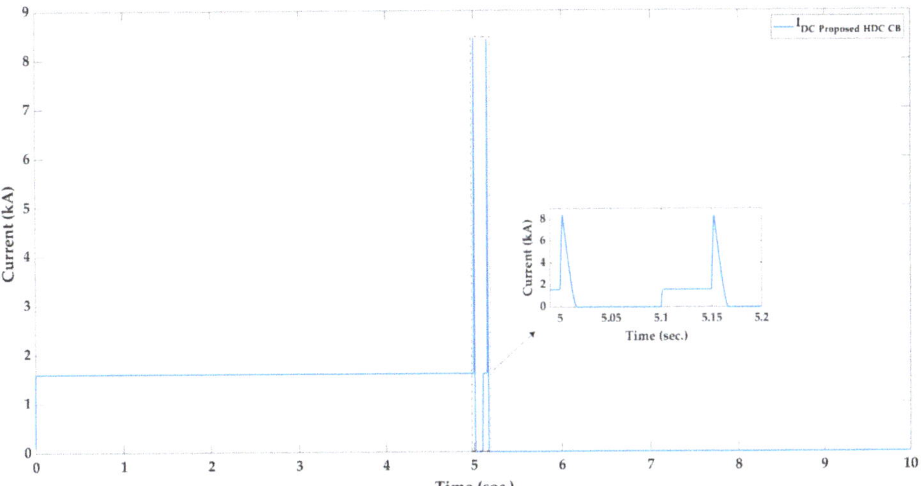

Figure 12. The DC current of the proposed HDCCB in reclosing and rebreaking test in the rated current interruption.

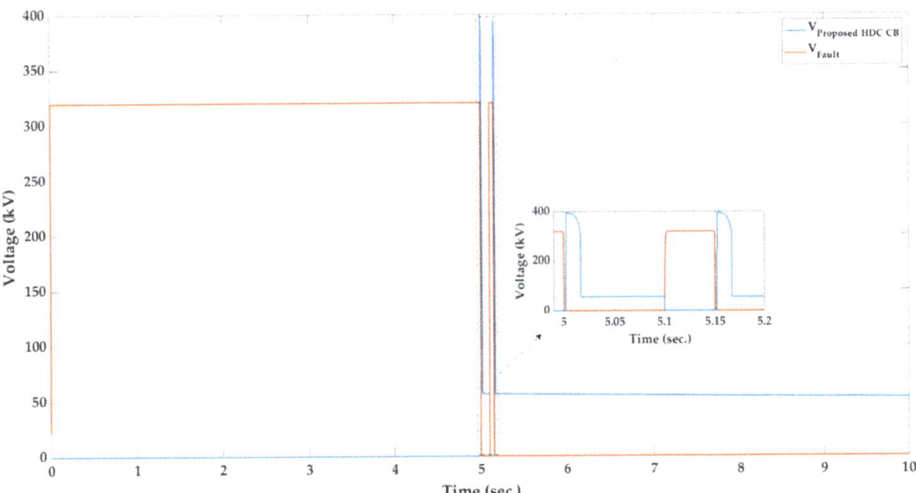

Figure 13. The CB voltage and voltage at the fault location in reclosing and rebreaking test in the rated current interruption.

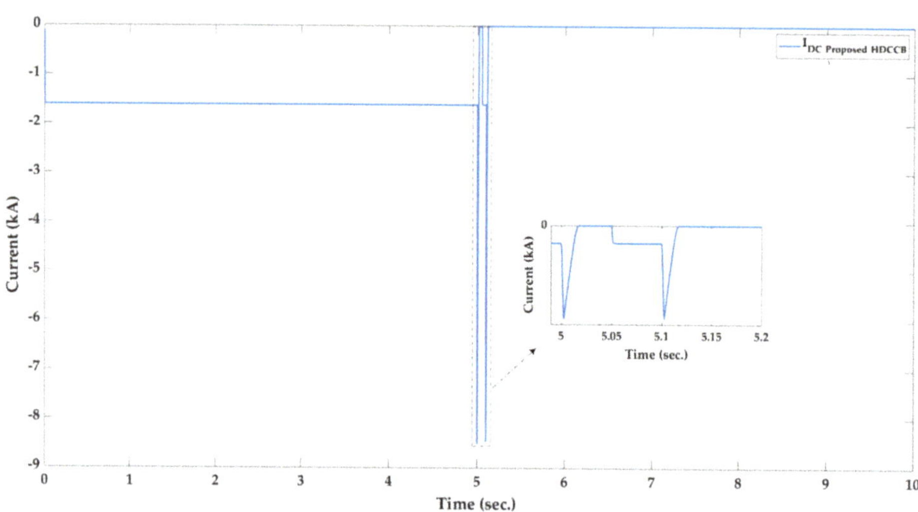

Figure 14. The DC current of the proposed HDCCB in reclosing and rebreaking tests in the reverse current interruption.

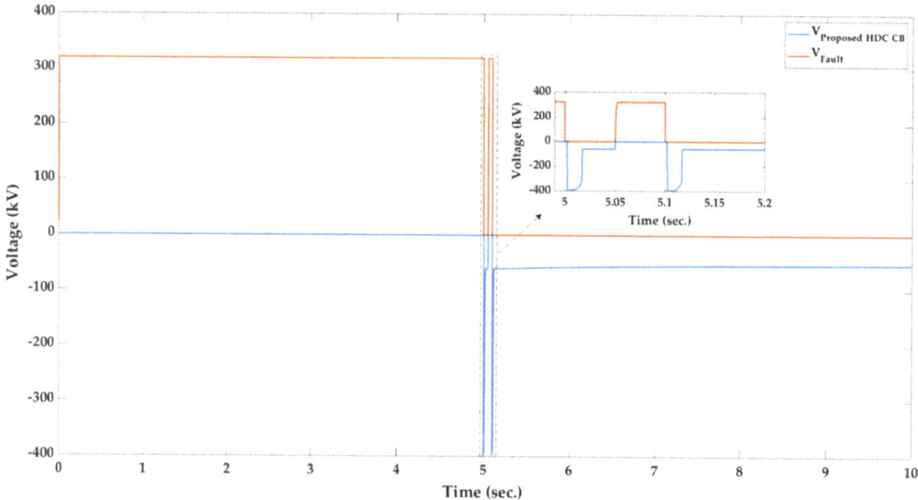

Figure 15. The CB voltage and voltage at the fault location in reclosing and rebreaking test in the reverse current interruption.

4.2. Comparison

To evaluate and compare the performance of the proposed HDCCB over the conventional DC CB in [22,29,30], replication is done, and proper matching is noticed in all cases. Figure 16 illustrates a comparison between the CB voltage responses of the proposed HDCCB and the conventional DC CB. As shown in Figure 16, the voltage response of the proposed HDCCB tracks the voltage of the conventional DC CB. However, due to the fact that the DC capacitors have some charges, the level of DC voltage of the proposed HDCCB after the fault is more than the level of voltage of the conventional DC CB.

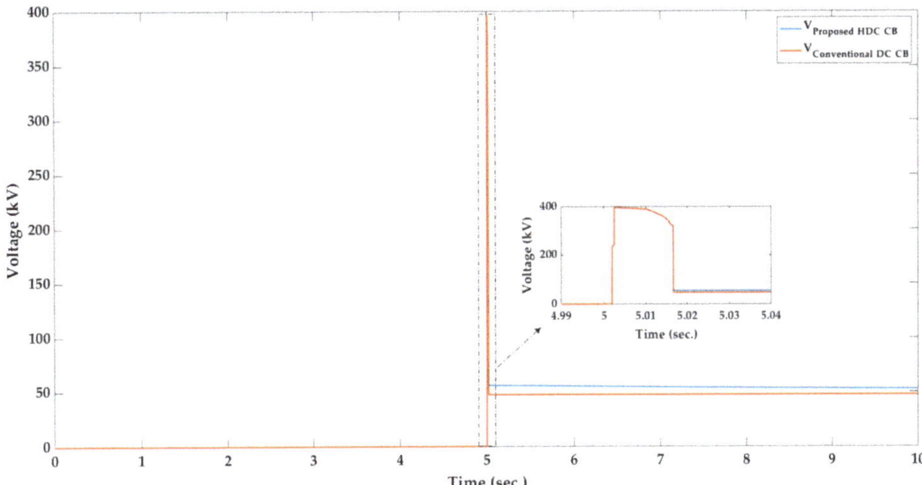

Figure 16. Comparison between the voltage response of the proposed HDCCB and the conventional DC CB after applying a short-circuit DC fault at 5 s.

Figure 17 demonstrates the comparison between the DC current response of the proposed HDCCB and the conventional DC CB after applying a short-circuit DC fault at 5 s. As shown in Figure 17, good matching is observed. The maximum breaking current capability of the proposed HDCCB and conventional DC CB are 8.4425 kA and 8.4207 kA, respectively. Thus, the difference between the maximum breaking current capability of the proposed HDCCB and conventional DC CB is 21.8 A.

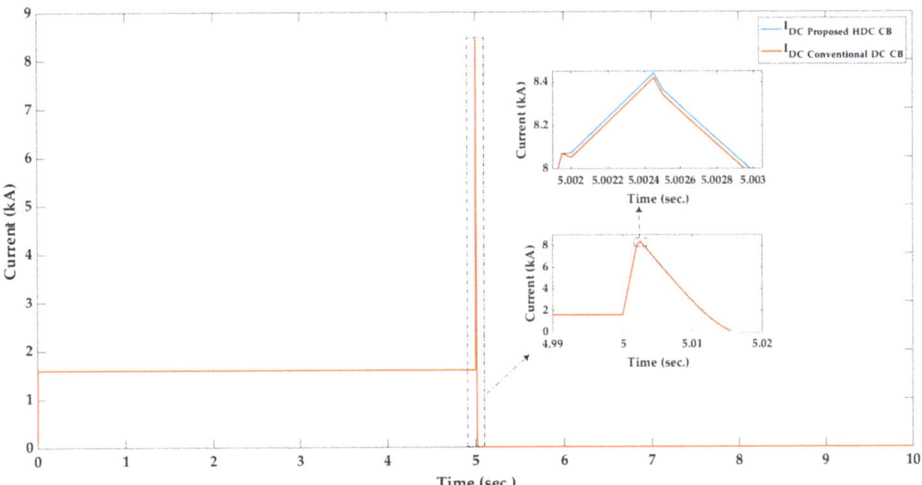

Figure 17. Comparison between the DC current response of the proposed HDCCB and the conventional DC CB after applying a short-circuit DC fault at 5 s.

The maximum dissipated energy of the surge arrestor in the DC CB 1 of the conventional DC CB is 12.6388 kJ, which is 17.9 J more than the corresponding value of the proposed HDCCB. However, because of the capacitor charge, the dissipated energy of the surge arrestor in the DC CB 2 of the conventional DC CB is 7.7952 kJ which is 16.1 J less than the corresponding value of the proposed

HDCCB. But, the total dissipated energy of the surge arrestor of the proposed HDCCB and the conventional DC CB are approximately the same.

5. Conclusions

This paper presents a new topology of a fast proactive Hybrid DC Breaker (HDCCB) to isolate the DC faults in Multi-Terminal VSC-HVDC (MT-HVDC) grids in case of fault current interruption, along with lowering the conduction losses and lowering the interruption time. All modes of operation of the proposed topology are studied. Simulation results verify that compared to the conventional DC CB, the proposed topology has lower interruption time, lower on-state switching losses, and higher breaking current capability. Due to the fact that MT-HVDC systems can share the power among the converter stations independently and in both directions, and considering the fast bidirectional fault current interruption, and reclosing and rebreaking capabilities, the proposed HDCCB can improve the overall performance of MT-HVDC systems and increase the reliability of the DC grids.

Author Contributions: F.M. was responsible for methodology, collecting resources, data analysis, writing—original draft preparation, and writing—review and editing. G.-A.N. and M.S. were responsible for the supervision, and writing—review and editing.

Funding: M. Saif acknowledges the financial support of Natural Sciences and Engineering Research Council (NSERC) of Canada, as well as National Science Foundation of China under grant number 61873144.

Conflicts of Interest: The authors declare no conflicts of interest.

References

1. Mohammadi, F.; Nazri, G.A.; Saif, M. A Fast Fault Detection and Identification Approach in Power Distribution Systems. In Proceedings of the IEEE 5th International Conference on Power Generation Systems and Renewable Energy Technologies (PGSRET), Istanbul, Turkey, 26–27 August 2019.
2. Jiang, H.; Zhang, J.J.; Gao, W.; Wu, Z. Fault Detection, Identification, and Location in Smart Grid Based on Data-Driven Computational Methods. *IEEE Trans. Smart Grid* **2014**, *5*, 2947–2956.
3. Mohammadi, F.; Zheng, C. Stability Analysis of Electric Power System. In Proceedings of the 4th National Conference on Technology in Electrical and Computer Engineering, Tehran, Iran, 27 December 2018.
4. Bucher, M.K.; Franck, C.M. Analytic Approximation of Fault Current Contributions from Capacitive Components in HVDC Cable Networks. *IEEE Trans. Power Deliv.* **2015**, *30*, 74–81. [CrossRef]
5. Shukla, A.; Demetriades, G.D. A Survey on Hybrid Circuit-Breaker Topologies. *IEEE Trans. Power Deliv.* **2015**, *30*, 627–641. [CrossRef]
6. Hassanpoor, A.; Häfner, J.; Jacobson, B. Technical Assessment of Load Commutation Switch in Hybrid HVDC Breaker. *IEEE Trans. Power Electron.* **2015**, *30*, 5393–5400. [CrossRef]
7. Belda, N.A.; Smeets, R.P.P. Test Circuits for HVDC Circuit Breakers. *IEEE Trans. Power Deliv.* **2017**, *32*, 285–293. [CrossRef]
8. Kontos, E.; Pinto, R.T.; Rodrigues, S.; Bauer, P. Impact of HVDC transmission system topology on multiterminal DC network faults. *IEEE Trans. Power Deliv.* **2015**, *30*, 844–852. [CrossRef]
9. Sneath, J.; Rajapakse, A.D. Fault Detection and Interruption in an Earthed HVDC Grid using ROCOV and Hybrid DC Breakers. *IEEE Trans. Power Deliv.* **2016**, *31*, 973–981. [CrossRef]
10. Ahmed, N.; Ängquist, L.; Mehmood, S.; Antonopoulos, A.; Harnefors, L.; Norrga, S.; Nee, H.P. Efficient Modeling of an MMC-Based Multiterminal DC System Employing Hybrid HVDC Breakers. In Proceedings of the IEEE Power and Energy Society General Meeting (PESGM), Boston, MA, USA, 17–21 July 2016.
11. Jovcic, D.; Taherbaneh, M.; Taisne, J.P.; Nguefeu, S. Offshore DC Grids as an Interconnection of Radial Systems: Protection and Control Aspects. *IEEE Trans. Smart Grid* **2015**, *6*, 903–910. [CrossRef]
12. Cwikowski, O.; Barnes, M.; Shuttleworth, R.; Chang, B. Analysis and simulation of the proactive hybrid circuit breaker. In Proceedings of the 2015 IEEE 11th International Conference on Power Electronics and Drive Systems, Sydney, Australia, 9–12 June 2015.
13. Wang, Y.; Marquardt, R. Future HVDC-Grids Employing Modular Multilevel Converters and Hybrid DC-Breakers. In Proceedings of the 15th European Conference on Power Electronics and Applications (EPE), Lille, France, 2–6 September 2013.

14. Wang, Y.; Marquardt, R. A Fast Switching, Scalable DC-Breaker for Meshed HVDC-SuperGrids. In Proceedings of the International Exhibition and Conference for Power Electronics, Intelligent Motion, Renewable Energy and Energy Management, Nuremberg, Germany, 20–22 May 2014.
15. Mokhberdoran, A.; Carvalho, A.; Leite, H.; Silva, N. A Review on HVDC Circuit Breakers. In Proceedings of the 3rd Renewable Power Generation Conference, Naples, Italy, 24–25 September 2014.
16. Guo, Q.; Yoon, M.; Park, J.; Jang, G. Novel Topology of DC Circuit Breaker and Current Interruption in HVDC Networks. *IEEJ Trans. Electr. Electron. Eng.* **2017**, *12*, 465–473. [CrossRef]
17. Tokoyoda, S.; Sato, M.; Kamei, K.; Yoshida, D.; Miyashita, M.; Kikuchi, K.; Ito, H. High Frequency Interruption Characteristics of VCB and Its Application to High Voltage DC Circuit Breaker. In Proceedings of the 3rd International Conference on Electric Power Equipment—Switching Technology, Busan, South Korea, 25–28 October 2015.
18. Corzine, K.A. A New-Coupled-Inductor Circuit Breaker for DC Applications. *IEEE Trans. Power Electron.* **2017**, *32*, 1411–1418. [CrossRef]
19. Keshavarzi, D.; Ghanbari, T.; Farjah, E. A Z-Source-Based Bidirectional DC Circuit Breaker with Fault Current Limitation and Interruption Capabilities. *IEEE Trans. Power Electron.* **2017**, *32*, 6813–6822. [CrossRef]
20. Maqsood, A.; Overstreet, A.; Corzine, K.A. Modified Z-Source DC Circuit Breaker Topologies. *IEEE Trans. Power Electron.* **2016**, *31*, 7394–7403. [CrossRef]
21. Mokhberdoran, A.; Carvalho, A.; Silva, N.; Leite, H.; Carrapatoso, A. A New Topology of Fast Solid-State HVDC Circuit Breaker for Offshore Wind Integration Applications. In Proceedings of the 17th European Conference on Power Electronics and Applications, Geneva, Switzerland, 8–10 September 2015.
22. Häfner, J.; Jacobson, B. Proactive Hybrid HVDC Breakers—A Key Innovation for Reliable HVDC Grids. In Proceedings of the CIGRÉ International Symposium: Electric Power System of the Future—Integrating Supergrids and Microgrids, Bologna, Italy, 13–15 September 2011.
23. Lin, W.; Jovcic, D.; Nguefeu, S.; Saad, H. Modelling of High-Power Hybrid DC Circuit Breaker for Grid-Level Studies. *IET Power Electron.* **2016**, *9*, 237–246. [CrossRef]
24. Liu, G.; Xu, F.; Xu, Z.; Zhang, Z.; Tang, G. Assembly HVDC Breaker for HVDC Grids With Modular Multilevel Converters. *IEEE Trans. Power Electron.* **2017**, *32*, 931–941. [CrossRef]
25. Feng, L.; Gou, R.; Yang, X.; Wang, F.; Zhuo, F.; Shi, S. A 320kV Hybrid HVDC Circuit Breaker Based on Thyristors Forced Current Zero Technique. In Proceedings of the IEEE Applied Power Electronics Conference and Exposition, Tampa, FL, USA, 26–30 March 2017.
26. Kim, B.C.; Chung, Y.H.; Hwang, H.D.; Mok, H.S. Comparison of Inverse Current Injecting HVDC Circuit Breaker. In Proceedings of the 3rd International Conference on Electric Power Equipment—Switching Technology, Busan, South Korea, 25–28 October 2015.
27. Mohammadi, F. Power Management Strategy in Multi-Terminal VSC-HVDC System. In Proceedings of the 4th National Conference on Applied Research in Electrical, Mechanical Computer and IT Engineering, Tehran, Iran, 4 October 2018.
28. Mohammadi, F.; Nazri, G.A.; Saif, M. A Bidirectional Power Charging Control Strategy for Plug-in Hybrid Electric Vehicles. *Sustainability* **2019**, *11*, 4317. [CrossRef]
29. Callavik, M.; Blomberg, A.; Häfner, J.; Jacobson, B. The Hybrid HVDC Breaker, an Innovation Breakthrough Enabling Reliable HVDC Grids. *ABB Grid Syst. Tech. Pap.* **2012**, *361*, 143–152.
30. Liu, W.; Liu, F.; Zhuang, Y.; Zha, X.; Chen, C.; Yu, T. A Multiport Circuit Breaker-Based Multiterminal DC System Fault Protection. *IEEE J. Emerg. Sel. Top. Power Electron.* **2019**, *7*, 118–128. [CrossRef]

Article

Optimal Power Control of Inverter-Based Distributed Generations in Grid-Connected Microgrid

Mohamed A. Hassan [1,*], Muhammed Y. Worku [1] and Mohamed A. Abido [2,3]

1 Center for Engineering Research, Research Institute, King Fahd University of Petroleum & Minerals, 31261 Dhahran, Saudi Arabia; muhammedw@kfupm.edu.sa

2 Electrical Engineering Department, King Fahd University of Petroleum & Minerals, Dhahran 31261, Saudi Arabia; mabido@kfupm.edu.sa

3 Senior Researcher at K. A. CARE Energy Research & Innovation Center, Dhahran 31261, Saudi Arabia

* Correspondence: mhassan@kfupm.edu.sa; Tel.: +966-13-860-7332

Received: 9 September 2019; Accepted: 16 October 2019; Published: 21 October 2019

Abstract: Distributed generation (DG) units are utilized to feed their closed loads in the autonomous microgrid. While in the grid-connected microgrid, they are integrated to support the utility by their required real and reactive powers. To achieve this goal, these integrated DGs must be controlled well. In this paper, an optimal PQ control scheme is proposed to control and share a predefined injected real and reactive powers of the inverter based DGs. The control problem is optimally designed and investigated to search for the optimal controller parameters by minimizing the error between the reference and calculated powers using particle swarm optimization (PSO). Microgrid containing inverter-based DG, PLL, coupling inductance, LC filter, power and current controllers is implemented on MATLAB. Two microgrid cases with different structure are studied and discussed. In both cases, the microgrid performance is investigated under different disturbances such as three-phase fault and step changes. The simulation results show that the proposed optimal control improves the microgrid dynamic stability. Additionally, the considered microgrids are implemented on real time digital simulator (RTDS). The experimental results verify the effectiveness and tracking capability of the proposed controllers and show close agreement with the simulation results. Finally, the comparison with the literature confirms the effectiveness of the proposed control scheme.

Keywords: distributed generation; dynamic stability; microgrid; and PQ Control

---

1. Introduction

With the fast-growing greenhouse gas emissions and other environmental issues, distributed generations (DGs) are rapidly connected to the electricity network [1,2]. Connecting different distributed energy resources (DER) with a group of loads is defined as a microgrid [3]. It acts as a single controllable entity with respect to the grid where it can be connected or disconnected from the grid to operate in both grid-connected and island modes respectively [3]. Different renewable sources and microgrid objectives such as stability, reliability resource penetration, AC and DC analysis, sustainability analysis, controlling voltage source converter (VSC) and DG integration are recently discussed extensively [4–8]. Integrating DGs with the grid can solve several typical problems of conventional AC network such as energy security and cost saving [2]. Microgrid capability to inject power to the grid while maintaining the system stability after getting disturbed is considered as one of the microgrid challenges [9,10]. Different control techniques such as active–reactive power control (PQ control), active power–voltage control (PV control) and voltage–frequency control (VF control) are utilized to control the DG units and achieve the required goals [11,12]. PQ scheme is used to control the exchanged real and reactive powers between the DG and grid [11]. Vf control is employed to keep the inverter voltage at constant value and return the frequency to its nominal value after getting real power disturbance [12]. For integrating inverter-based DG with the power system, several PQ control schemes such as hysteresis, dead-beat (DB) controllers, proportional–integral (PI) controllers and proportional–resonant (PR) have been proposed [13–19]. Hysteresis control is simple and has fast responses, but the output current contains high ripples leading to poor current quality and finding some difficulties to design the output filter [13]. DB predictive control is widely used because it offers high performance for current-controlled DGs. Nevertheless, it is quite complicated and sensitive to system parameters [14]. If the target is to compensate multiple harmonics behind eliminate the steady-state error and regulate the sinusoidal signals, PR control scheme in the stationary (α, β) reference frame is popular [15]. However, to maintain good performance, the resonant frequency and the varying grid frequency should be identical [16]. PI controller has many advantages such as instantaneous control, better wave shaping and fixed inverter switching frequency resulting in known harmonics [17–19]. PQ controller usually adopts double loop controls [20]. Based on power target, the outer power loop produces the reference current while the current inner loop plays the role of fine-tuning [20]. The current control is implemented in the rotating (synchronous) *dq* reference frame because the synchronous frame controller can eliminate steady state error and has fast transient response.

In the literature, several PQ control techniques have been presented to control the injected powers of the DGs in the grid-connected microgrid [21–26]. In our previous work [21], a power controller was implemented to calculate the *dq* reference currents using the *dq* output voltages and reference real and reactive powers. Utilizing Newton–Raphson-based parameter estimation and feed forward control approaches, a robust servomechanism voltage controller and a discrete-time sliding mode current controller were used to control the DG power flow in the grid-connected mode [22]. For a single-phase grid-connected fuel cell system, a second order generalized integrator to control the active and reactive powers was presented [23]. Based on six control degrees, an individual-phase decoupled PQ controller was anticipated for a three-phase VSC [24]. In grid-connected microgrid, utilizing maximum power point tracking (MPPT), a PQ control method was proposed to control the power of the solar photovoltaic and battery storage [25]. PQ control was used in the load-following mode and PV control was utilized in maximum power point tracking mode to control a solar photovoltaic in distribution systems [26]. Unfortunately, all the previous presented work suffers from the bad performance especially under dynamic loads and generation variations since they are relying on deeply empirical engineering rules for designing the multivariable parameters of the PQ controllers [26]. Designing an optimal PQ controller is essentially to overcome the aforementioned problems and to solve the constrained optimization problem [26]. Recently, optimal control using swarm algorithms and popular evolutionary has been effectively utilized in power systems and power converters [27–32]. Moreover, designing an optimal PQ controller has been reported in a few works [26,30]. In [26], an optimal active and reactive power control was developed for a three-phase grid-connected inverter in a microgrid by using an adaptive population-based extremal optimization algorithm (APEO). In the grid-connected microgrid, a particle swarm optimization (PSO)-based PQ control technique under variable loads conditions was proposed [30]. This important work has confirmed the importance of PSO in the automatic tuning of PQ control parameters for optimized operation during abrupt load changes. However, this work did not optimize the controller parameters of the two current controllers in the designed control system. Therefore, the problem may not be considered as an incomplete optimization process for designing PQ controllers [26]. Moreover, they did not consider the optimization of the filter components which they have a great effect on the microgrid stability [3]. Additionally, the microgrid will be under more stress and the controller design needs to be more accurate when we have a generation disturbance not a load disturbance as presented in [26].

In this paper, an efficient PI power controller is proposed to regulate the predefined injected real and reactive powers to the grid. The control problem is optimally designed based on minimizing the error between the calculated and the injected powers to get the optimal controller parameters. Particle swarm optimization (PSO) is employed to design the controller parameters and LC filter components. Different microgrid structures are implemented and examined in MATLAB. Firstly, the optimal proposed controller is designed to control the injected real and reactive powers of one inverter-based DG. Secondly, the optimal proposed controller is designed for two different rated inverter-based DG units to share their injected powers to the grid. Sever disturbances such as step up/down changes of real and reactive injected powers, three-phase fault and losing DG unit are applied to investigate the proposed controller effectiveness and to ensure the system stability after getting disturbed. Additionally, to validate the usefulness of the proposed controller, the considered microgrid is implemented on real time digital simulator (RTDS). To confirm the effectiveness of the proposed optimal control scheme, it is compared with the exiting work in [21] through extensive simulation and experiments under various disturbances. The results confirm the superiority of the proposed control strategy in providing a fast, accurate and decoupled power control with a lower AC current distortion.

The major contributions of this work are described as follows:

(1) A new optimal PQ control scheme is proposed for inverter-based grid-connected microgrid to improve the microgrid dynamic stability.
(2) The proposed scheme is compared with the exciting control scheme to validate the proposed controller robustness. The superiority of the proposed control is confirmed using both MATLAB simulation and RTDS experimental results for an inverter-based grid-connected microgrid.
(3) The proposed controller has been verified for a two inverter-based grid-connected microgrid.
(4) To the best of the authors' knowledge, an optimal PQ control technique is firstly implemented in real time digital simulator (RTDS) to control the injected real and reactive powers of the inverter-based DGs in the grid-connected microgrid.
(5) The superiority of the proposed method is demonstrated by experimental results using RTDS.

2. System Description

Microgrid is defined as one or more DG units connected closely with loads through coupling inductance and LC filter [3]. DGs such as PV generate DC power therefore; they usually utilize an inverter to convert the DC power to AC. Microgrid operates in two different modes; island and grid-connected modes. In the island or autonomous mode, maintaining the voltage and frequency of the system and supporting the required active and reactive powers is the main task [27–31]. In the grid-connected mode, the main target is to control the delivered DGs powers into the grid. In this paper, to control a predefined real and reactive powers to the grid, two cases are considered. In the first case, an inverter-based DG is delivering power to the grid through a coupling inductance and LC filter as shown in Figure 1. While in the second case, two inverter-based DG units are sharing their powers to the grid at the point of common coupling (PCC) through a coupling inductance and LC filter as shown in Figure 2.

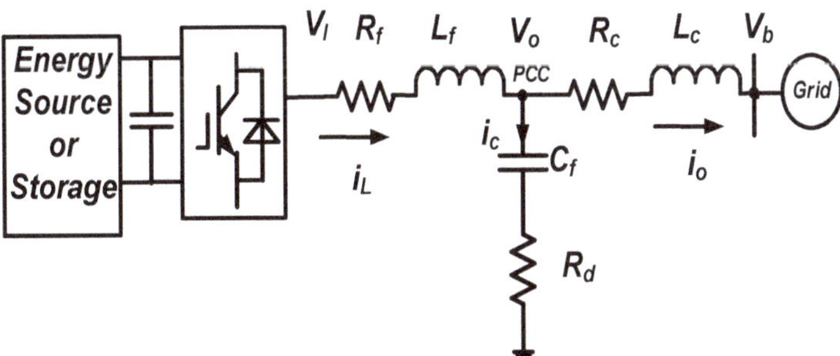

Figure 1. Grid-connected microgrid (One distributed generation's (DGs) case).

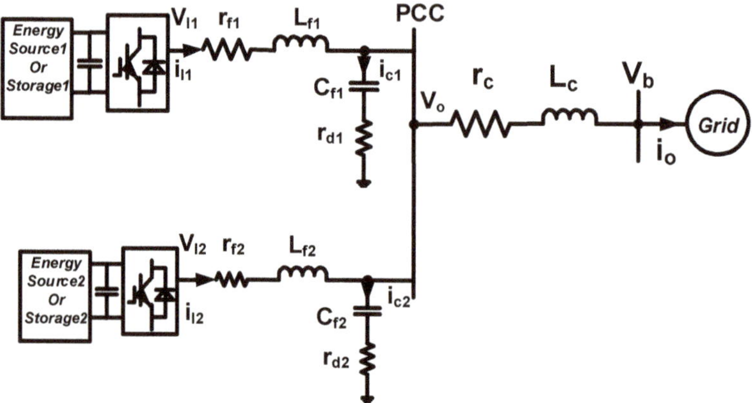

Figure 2. Grid-connected microgrid (Two DGs case).

From Figures 1 and 2:

C_f, L_f and R_f are the capacitance, inductance and resistance of the LC filter,
L_c and R_c are the inductance and resistance of the coupling inductor,
R_d is the damping resistance,
i_L is the coupling inductor current,
i_o is the inverter output current,
i_c is the capacitor current,
V_I is the inverter output voltage,
V_o is the PCC voltage,
V_g is the grid voltage.
C_{f1}, L_{f1} and r_{f1} are the capacitance, inductance and resistance of the LC filter for DG1,
C_{f2}, L_{f2} and r_{f2} are the capacitance, inductance and resistance of the LC filter for DG2,
i_{L1} and i_{L2} are the coupling inductor currents of DG1 and DG2 respectively,
i_{c1} and i_{c1} are the capacitor currents of DG1 and DG2 respectively,
V_{I1} and V_{I2} are the inverter output voltages of DG1 and DG2 respectively,
r_{d1} and r_{d2} are the damping resistances.

In both cases, it is worth mentioning that each DG inverter is assumed to be connected to a constant DC power source, so there no need to regulate the DC-link voltage otherwise, a controller should be introduced to regulate the DC-link voltage [13]. Meanwhile, our main objective in this paper is to study the AC side dynamic performance of the inverter-based DG in the grid-connected mode.

3. Proposed Methodology

Depending on the grid demand, the DG inverter is controlled to inject specific amount of power. An optimal PQ control scheme illustrated in Figure 3 is proposed adopting double loop controls to improve the dynamic performance of the grid-connected microgrid. The proposed PI power controller is firstly implemented to produce the reference current signal based on injected power. Secondly, the current control loop considers several aspects such as providing injected three-phase balanced currents, obtaining high power quality and overcoming the nonlinearities coming from the interaction between inverter switching and external disturbances [21].

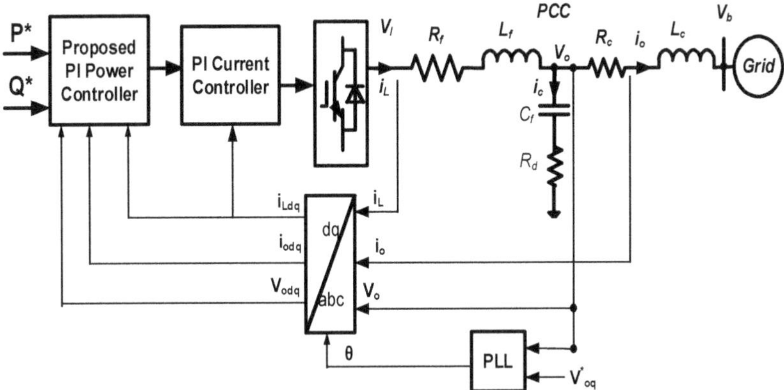

Figure 3. Proposed microgrid controller.

The proposed methodology can be summarized as follows. Firstly, the optimization process will be initiated where PSO generates random controller parameters. These random parameters are used to calculate the objective function. The minimum objective function and its corresponding optimal parameters will be saved. Secondly, the inverter output voltage and current are measured and converted into *dq* forms. Thirdly, the real and reactive powers are calculated using the output voltage and current. Fourthly, the optimal controller parameters of the power controller are used to generate the reference currents by comparing the calculated powers with reference powers. Finally, the optimal controller parameters of the current controller are employed to generate the reference voltages to fire the inverters by comparing the reference currents with measured currents. These steps will be explained in detail in the next sections.

Firstly, the three-phase output voltage v_o and current i_o are measured at PCC as shown in Figures 1 and 2. Then the measured output voltage v_o is converted to the *dq* components using the transformation angle θ as given in Equation (1). Similarly, the *dq* components of the output current i_o can be obtained. The angle θ is obtained using the phase locked loop (PLL) shown in Figure 4a,b. Considered as one of the most common synchronization methods, PLL is used to extract the phase angle θ of the grid voltage and keep the output signal synchronized with the frequency and phase of the reference input signal [31].

$$\begin{pmatrix} v_{od} \\ v_{oq} \\ v_{oo} \end{pmatrix} = \sqrt{\frac{2}{3}} \begin{pmatrix} \cos\theta & \cos\left(\theta - \frac{2\pi}{3}\right) & \cos\left(\theta + \frac{2\pi}{3}\right) \\ -\sin\theta & -\sin\left(\theta - \frac{2\pi}{3}\right) & -\sin\left(\theta + \frac{2\pi}{3}\right) \\ \frac{1}{\sqrt{2}} & \frac{1}{\sqrt{2}} & \frac{1}{\sqrt{2}} \end{pmatrix} \begin{pmatrix} v_{oa} \\ v_{ob} \\ v_{oc} \end{pmatrix} \tag{1}$$

$$\omega = k_P^{PLL}(v_{oq} - v_{oq}^*) + k_I^{PLL} \int (v_{oq} - v_{oq}^*)dt \tag{2}$$

$$\theta = \int (\omega - \omega_{ref})dt + \theta(0) \tag{3}$$

where ω and ω_{ref} are the nominal and reference frequencies, v_{oq} and v^*_{oq} are the q components of the inverter output and reference voltages v_o and v^*_o, $k_P{}^{PLL}$, $k_I{}^{PLL}$ are the PI controller parameters of the PLL.

Secondly, the *dq* components of the output voltage and current are used to calculate the real and reactive powers (P_{cal} and Q_{cal}) as given in [3].

$$P_{\text{cal}} = v_{od}i_{od} + v_{oq}i_{oq} \tag{4}$$

$$Q_{\text{cal}} = v_{od}i_{oq} - v_{oq}i_{od} \tag{5}$$

where v_{od} and v_{oq} are the dq components of the inverter output voltage v_o, i_{od} and i_{oq} are the dq components of the inverter output current i_o.

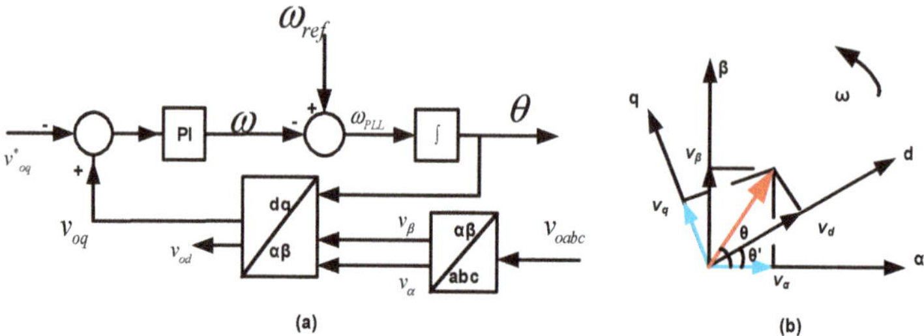

Figure 4. Phase locked loop (PLL) model.

Thirdly, by comparing the reference real and reactive powers (P^* and Q^*)and the calculated real and reactive powers (P_{cal} and Q_{cal}) respectively, the proposed PI power controller is deployed to produce the dq components of the output reference currents (i^*_{od} and i^*_{oq}) as given in Equations (6) and (7). The injected powers to the grid could track the reference power. Note that real and reactive power can be controlled independently because of the decoupling of the reference current (6) and (7) as shown in Figure 5.

$$i^*_{od} = k_{pp}(P^* - P_{cal}) + k_{ip}\int (P^* - P_{cal})dt \tag{6}$$

$$i^*_{oq} = k_{pq}(Q^* - Q_{cal}) + k_{iq}\int (Q^* - Q_{cal})dt \tag{7}$$

where k_{pp}, k_{ip}, k_{pq} and k_{iq} are the PI power controller parameters.

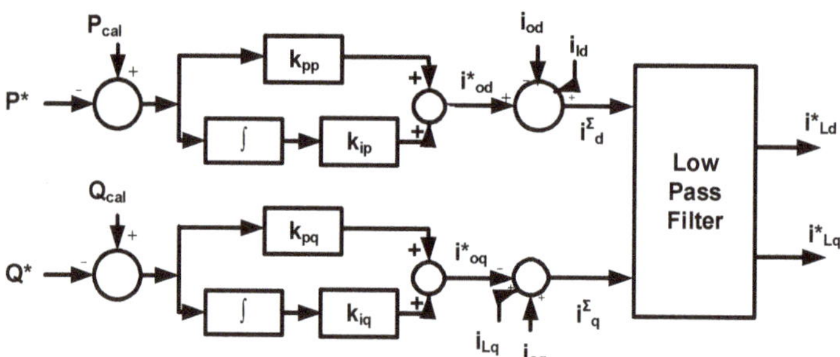

Figure 5. Proposed power controller.

The inverter is controlled to inject the coupling inductor current i_L not the inverter output current i_o as shown in Figures 1 and 2. Therefore, to obtain the coupling inductor reference current $i^\Sigma{}_L$ in the dq frame, the dq components of the output reference currents i^*_o is added to the dq components of the capacitor current i_c as given in Equations (8) and (9) and shown in Figure 5.

$$i^\Sigma_d = i^*_{od} + i_{cd} = i^*_{od} + (i_{Ld} - i_{od}) \tag{8}$$

$$i^\Sigma_q = i^*_{oq} + i_{cq} = i^*_{oq} + (i_{Lq} - i_{oq}) \tag{9}$$

Then the fundamental reference currents i^*_{Ld} and i^*_{Lq} can be obtained using a low-pass filter [21].

$$i^*_{Ld} = \frac{\omega_c^2}{s^2 + \sqrt{2}s\omega_c + \omega_c^2} i^\Sigma_d \tag{10}$$

$$i^*_{Lq} = \frac{\omega_c^2}{s^2 + \sqrt{2}s\omega_c + \omega_c^2} i_q^{\Sigma} \quad (11)$$

where ω_c is the cut-off frequency of the low-pass filter.

Fourthly, the PI current controller shown in Figure 6 is used to obtain the *dq* components of the reference voltage v^*_l targeting zero steady state error and compensating both inductor non-idealities and inverter switching nonlinearities [21].

$$v^*_{ld} = v_{od} - \omega L_f i_{Ld} + k_P^d (i^*_{Ld} - i_{Ld}) + k_I^d \int (i^*_{Ld} - i_{Ld}) dt \quad (12)$$

$$v^*_{lq} = v_{oq} + \omega L_f i_{Lq} + k_P^q (i^*_{Lq} - i_{Lq}) + k_I^q \int (i^*_{Lq} - i_{Lq}) dt \quad (13)$$

where i_{Ld}, i_{Lq} are *dq* components of the coupling inductor current i_L, i^*_{Ld}, i^*_{Lq} are *dq* components of the reference controller current i^*_L, $k_P{}^d$, $k_I{}^d$, $k_P{}^q$, $k_I{}^q$ are PI current controller parameters.

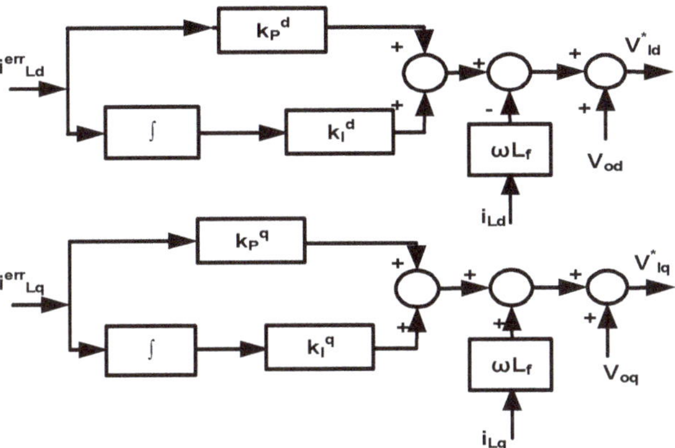

Figure 6. Current controller.

Additionally, the relationship between the PCC output voltage v_o and the inverter voltage v_l are given by Equations (14), (15), and (16) [32].

$$\begin{bmatrix} v_{la} \\ v_{lb} \\ v_{lc} \end{bmatrix} = R_f \begin{bmatrix} i_{la} \\ i_{lb} \\ i_{lc} \end{bmatrix} + L_f \frac{d}{dt} \begin{bmatrix} i_{la} \\ i_{lb} \\ i_{lc} \end{bmatrix} + \begin{bmatrix} v_{oa} \\ v_{ob} \\ v_{oc} \end{bmatrix} \quad (14)$$

$$v_{ld} = v_{od} + R_f i_{od} + L_f \frac{di_{od}}{dt} - \omega L_f i_{oq} \quad (15)$$

$$v_{lq} = v_{oq} + R_f i_{oq} + L_f \frac{di_{oq}}{dt} + \omega L_f i_{od} \quad (16)$$

To reduce the inverter switching frequency ripple, a low-pass filter is used. Additionally, a damping resistance is added to evade the possible resonance between this filter and the coupling inductance shown in Figure 1 [21]. Their models are given as follows:

$$v_{la} = i_{La} R_f + L_f \frac{di_{La}}{dt} + v_{Ca} + i_{Ca} R_d \quad (17)$$

$$v_{ba} = -i_{oa} R_c - L_c \frac{di_{oa}}{dt} + v_{Ca} + i_{Ca} R_d \quad (18)$$

$$\frac{dv_{Ca}}{dt} = \frac{1}{C_f}(i_{La} - i_{oa}) \tag{19}$$

where v_{La}, v_{Lb}, vLc are inverter output voltages.

4. Optimal Controller Design

Based on time domain simulation, the control problem is designed and formulated as an optimization problem where PSO is employed to minimize the proposed objective function J aiming to obtain the optimal controller and filter parameters [3].

$$J = \int_{t=0}^{t=t_{sim}} (P_{cal} - P^*)^2 .t\ dt \tag{20}$$

where t is added to ensure minimum settling time, t_{sim} the simulation time, and P_{cal} and P^* are the calculated and reference real power of the inverter-based DG respectively.

The problem constraints are the controller and filter parameters K= $[k_{pp}\ k_{ip}k_{pq}k_{iq}k_p{}^d\ k_i{}^d k_p{}^q k_i{}^q L_f C_f, R_d]^T$ bounded as follows:

$$K^{\min} \leq K \leq K^{\max} \tag{21}$$

where $k_{pp}\ k_{ip}k_{pq}$ and k_{iq} are the PI controller parameters of the proposed power controller while $k_p{}^d\ k_i{}^d k_p{}^q$ and $k_i{}^q$ are the PI controller parameters of the current controller. L_fa and Cf are the filter inductance and capacitance respectively. R_d is the damping resistance.

In 1995, PSO is a population based stochastic optimization method developed by Eberhart and Kennedy inspired by social behavior of bird flocking or fish schooling [33]. It is worth mentioning that PSO is used as an efficient tool for optimization that gives a balance between local and global search techniques. PSO advantages—like computational efficiency, simplicity, and robustness—will enhance the microgrid transient performance [3]. Using PSO, the best solution (candidate) of the population could be obtained by starting random particles selection and updating the generations inside this population. Ensuring the optimal solution convergence, the particles are moving in the search space trying to follow the optimum particles. In the minimization problem, at a given position, the highest fitness corresponds to the lowest value of the objective function at that position. At iteration (n + 1), the new position of each particle is obtained by Equation (22) as follows:

$$k_{n+1}^i = k_n^i + v_{n+1}^i \tag{22}$$

where $k^i{}_{n+1}$ is the position of particle i at iteration $n + 1$; kin is the position of particle i at iteration n; and $v^i{}_{n+1}$ is the corresponding velocity vector.

At each time step, the velocity of each particle is modified depending on both current velocity and current distance from the personal and global best positions as follows:

$$v_{n+1}^i = wv_n^i + c_1 r_1 p_{best} - k_n^i + c_2 r_2 (g_{best} - k_n^i) \tag{23}$$

where w is the inertia weight; $v^i{}_n$ is the velocity of particle i at iteration n; r_1 and r_2 are random numbers between 0 and 1; p_{best} is the best position found by particle i; g_{best} is the best position in the swarm at time n; and c_1 and c_2 are the "trust" parameters. Figure 7 shows the proposed PSO computational flow chart. The PSO steps are summarized in [7].

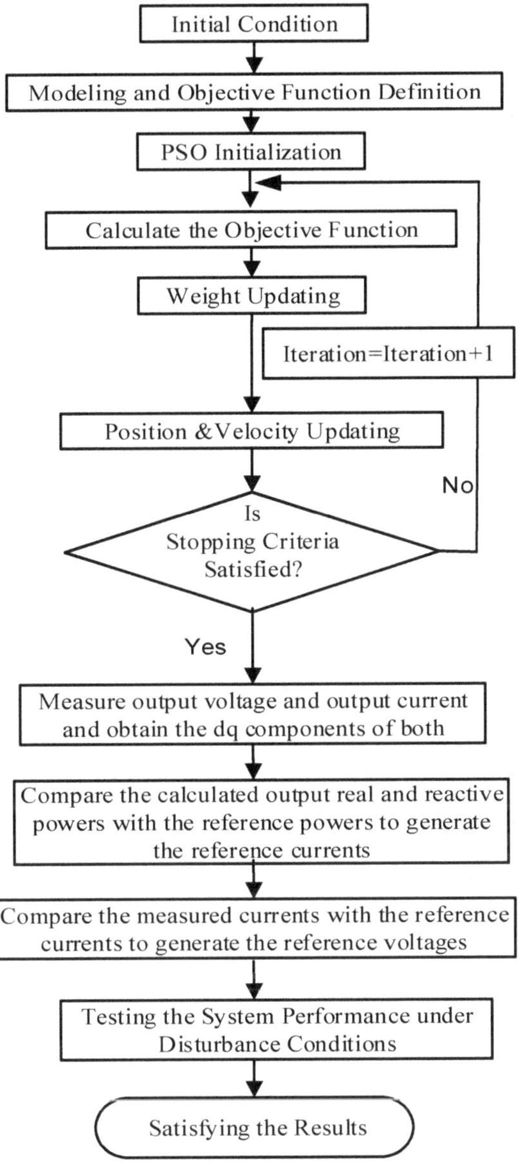

Figure 7. Proposed particle swarm optimization (PSO) computational flow chart.

5. Simulation Results and Discussion

In this work, to verify the effectiveness of the PQ control in the grid-connected microgrid, two cases have been simulated. Firstly, 5 kVA inverter-based DG is controlled to deliver a predefined real and reactive powers to the grid. Secondly, two different rated inverter-based DG units (5 kVA and 10 kVA) are sharing their injected controlled powers with the grid. Assuming an ideal source from the DG side, the DC bus dynamics is neglected. With the realization of high switching frequencies (4–10 kHz), the switching process of the inverter may also be neglected [14]. A simulation model for the proposed microgrid cases is built in a MATLAB based on the control strategy. The proportional gain k_p and integral gain k_i of the power and current controllers have been optimally tuned using PSO. Using the time domain simulation, the microgrid dynamic stability has been investigated and the proposed controller effectiveness has been evaluated under the following different disturbances:

1. Step change in the injected real power.
2. Step change in the injected reactive power.
3. Simultaneous step change in both injected real and reactive powers.
4. Three-phase fault at the PCC.

The optimal parameters for both cases are given respectively in Tables 1 and 2.

Table 1. Optimal parameters for one DG case.

PI Power Controller Parameters			
$k_{pp(Amp/Watt)}$	$k_{ip(Amp/Joule)}$	$k_{pq(Amp/Watt)}$	$k_{iq(Amp/Joule)}$
0.000737	5.03138	0.000737	5.03138
PI Current Controller Parameters			
k_i^d(Volt/Current. Sec)	k_p^q(Volt/ Amp)	k_i^q(Volt/Current. Sec)	k_p^d(Volt/ Amp)
649.54	8.87277	649.54	8.87277
Filter Parameters			
C_f (μF)	L_f (mH)		R_d (Ω)
10.4	2.176		10.6539

Table 2. Optimal parameters for two DGs case.

PI Power Controller Parameters				
	$k_{pp(Amp/Watt)}$	$k_{ip(Amp/Joule)}$	$k_{pq(Amp/Watt)}$	$k_{iq(Amp/Joule)}$
DG1	0.0009	5.0	0.0009	5.0
DG2	0.0008542	5.97056	0. 0008542	5.97056
PI Current Controller Parameters				
	k_i^d (Volt/Current. Sec)	k_p^q(Volt/ Amp)	k_i^q(Volt/Current. Sec)	k_p^d(Volt/ Amp)
DG1	606.375	10	606.375	10
DG2	632.285	5.1117	632.285	5.1117
Filter Parameters				
	C_f(μF)	L_f(mH)		R_d(Ω)
	11.0	5		10.0

5.1. One Inverter-Based DG Case

The proposed controller has been tested when a 5 kVA inverter-based DG connected to the grid. Firstly, the reference real power has been stepped down from 5 kW to 3 kW at t = 0.1 sec. Figure 8 shows the calculated and reference real and reactive powers for this disturbance. Both injected real and reactive powers are mostly following the reference powers. While Figure 9 depicts the calculated and reference real and reactive powers when the injected reactive power has been stepped down from 5 kVAR to 3 kVAR without any change in the injected real power. Both injected real and reactive powers are almost tracking the reference powers. It could be observed from the results that the controller has quick responses and track the references effectively. Additionally, the results show a reasonable steady state response at the beginning and even after clearing disturbance. (show as Tables 1 and 2).

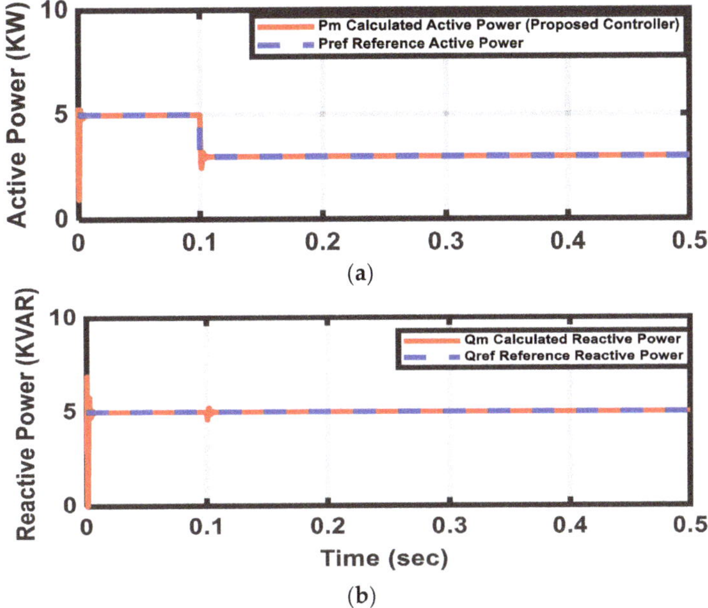

Figure 8. (**a**) Active power response at real power step down change. (**b**) Reactive power response at real power step down change.

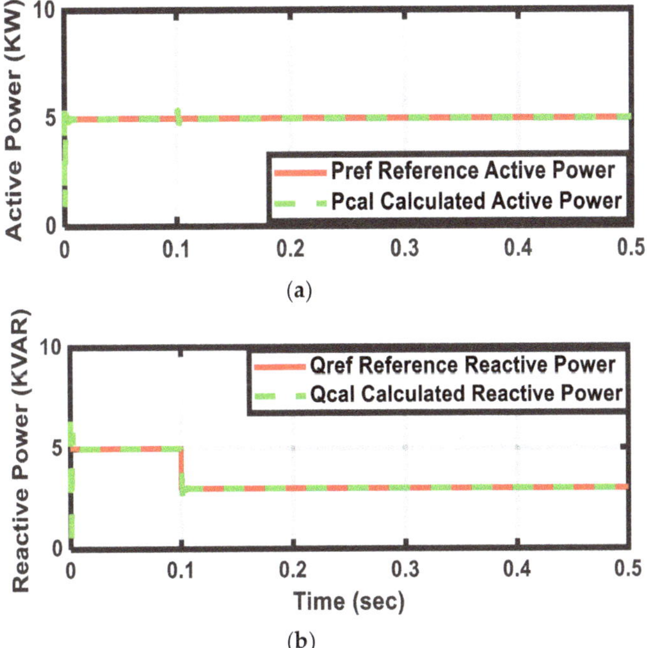

Figure 9. (**a**) Active power response at real power step down change. (**b**) Reactive power response at real power step down change.

Secondly, Figure 10 shows the microgrid dynamic response with step changes in real and reactive powers. Both active and reactive powers have been simultaneously stepped down from 5 kW to 3 kW and from 5 kVAR to 3 kVAR respectively at t = 0.1 sec. The results show how the proposed controller is working perfectly making the calculated powers track the reference powers perfectly.

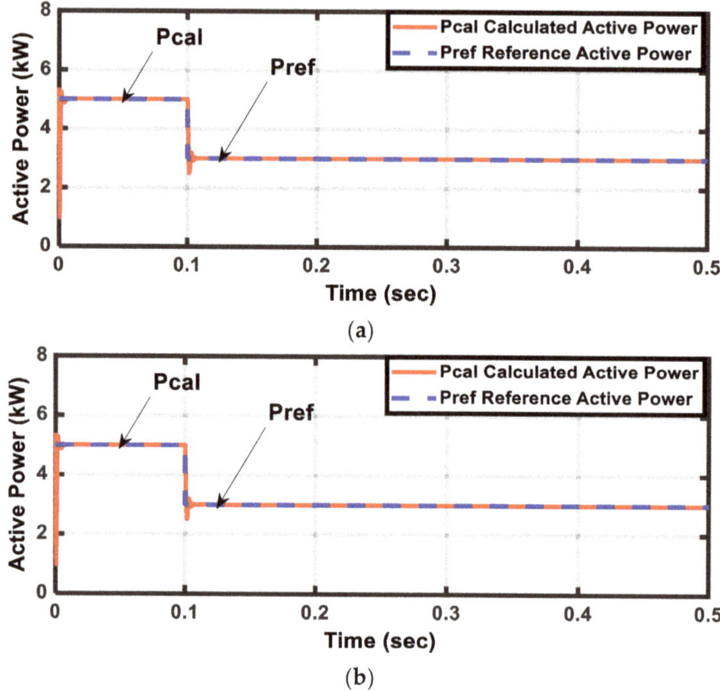

Figure 10. (**a**) Active power response at real and reactive powers step down changes. (**b**) Reactive power response at real and reactive powers step down changes.

Thirdly for further testing of the controller robustness, a three-phase fault disturbance has been applied at the PCC at t = 0.1 sec. Fault has been cleared at t = 0.4 sec. Figure 11 shows the dynamic responses of the three-phase output currents and their *dq* components where the system has been recovered from the fault. The proposed controller shows a very good transient performance. The calculated real and reactive powers are following the reference real and reactive powers in a good way under this fault disturbance.

Finally, to confirm the superiority of the proposed controller, a comparison between the proposed controller and the existing control scheme (power calculator presented in [21]) has been carried out. As it was mentioned before, the reference currents were calculated using the reference powers and measured output voltage in [21] while in the proposed method, an optimal power PI controller has been implemented to obtain the reference currents where the calculated powers were compared with the reference powers. Figures 12 and 13 show the dynamic response of the *dq* components of the inverter currents, and real and reactive powers when the injected real power stepped down from 5 kW to 3 kW at t = 0.1 sec. It can be shown from the results that the proposed controller shows better performance in terms of tracking the reference powers and steady state response. Figures 14 and 15 illustrate the dynamic response of the real and reactive powers for stepping up the injected reactive power from 3 kVAR to 5 kVAR and stepping down the injected real power from 5 kW to 3 kW at t = 0.1 sec. The results illustrate how the calculated powers follow perfectly the reference powers without any delay in the proposed controller while in the existing control scheme [21], there is a small delay when the calculated powers try to track the reference powers. Figures 16 and 17 depict the dynamic response of the *dq* components of the inverter currents, real and reactive powers for simultaneous step changes in both real and reactive powers from 3 kW to 5 kW and from 3 kVAR to 5 kVAR respectively at t = 0.1 sec. The system response due to a three-phase fault applied at the PCC is depicted in Figure 18. The results illustrate the superiority of the proposed controller. The response of the microgrid equipped with the proposed controller keep better and fast reference tracking in a good way.

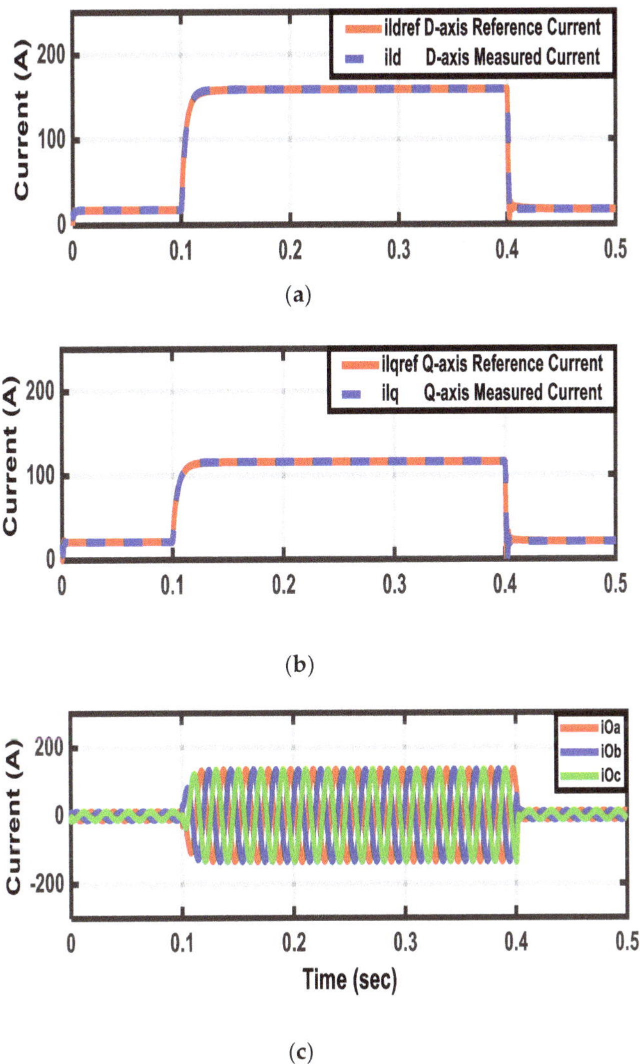

Figure 11. (**a**) D-axis current response at three-phase fault disturbance at PCC. (**b**) Q-axis current response at three-phase fault disturbance at PCC. (**c**) Three-phase current response at three-phase fault disturbance at PCC.

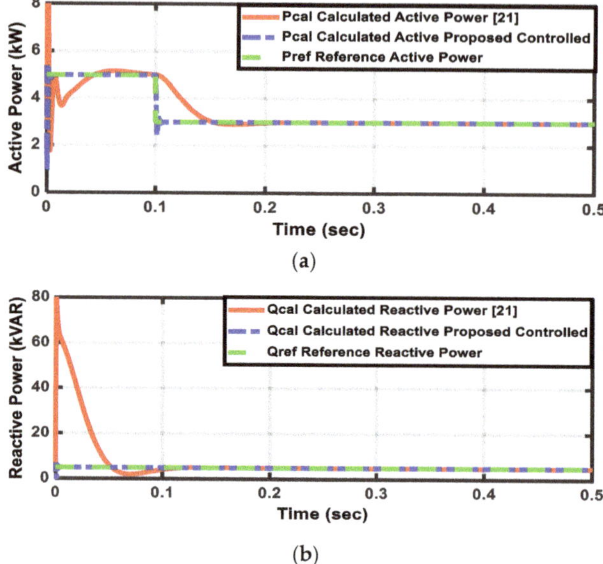

Figure 12. (**a**) Active power response at step down change in real power. (**b**) Reactive power response at step down change in real power.

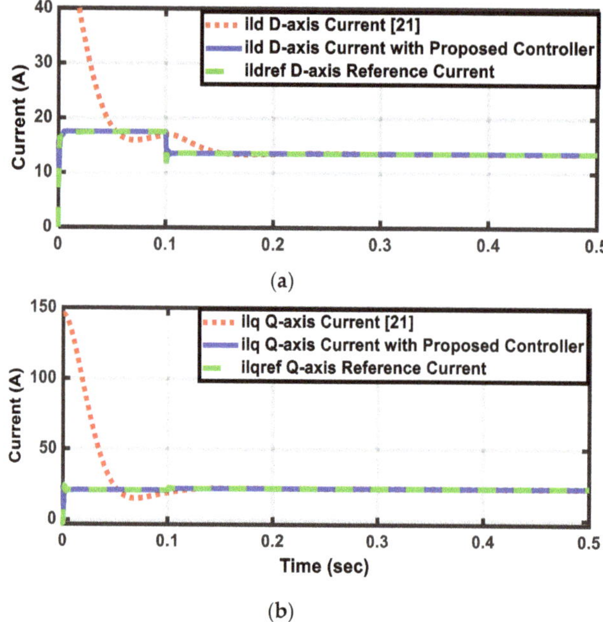

Figure 13. (**a**) D-axis current responses at step down change in real power. (**b**) Q-axis current responses at step down change in real power.

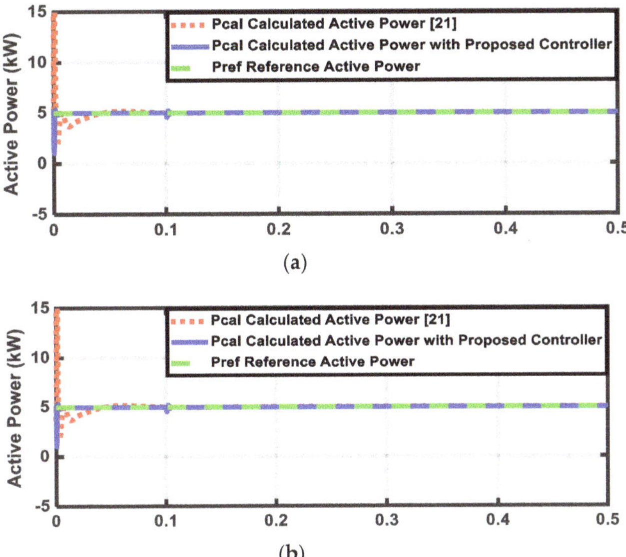

Figure 14. (**a**) Active power response at step up change in reactive power. (**b**) Reactive power response at step up change in reactive power.

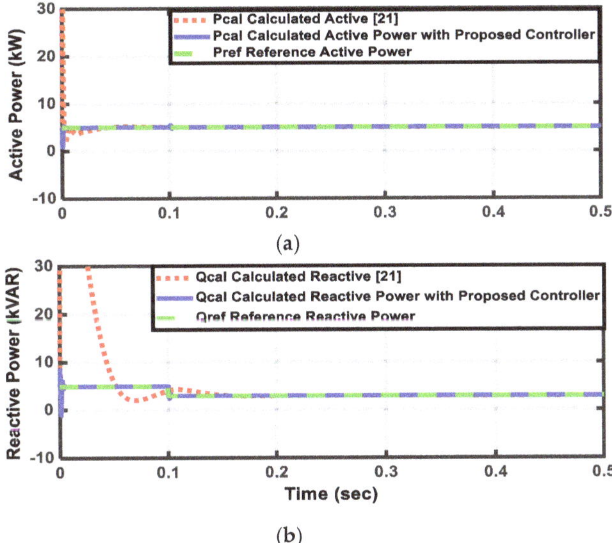

Figure 15. (a) Active power response at step down change in reactive power. (**b**) Reactive power response at step down change in reactive power.

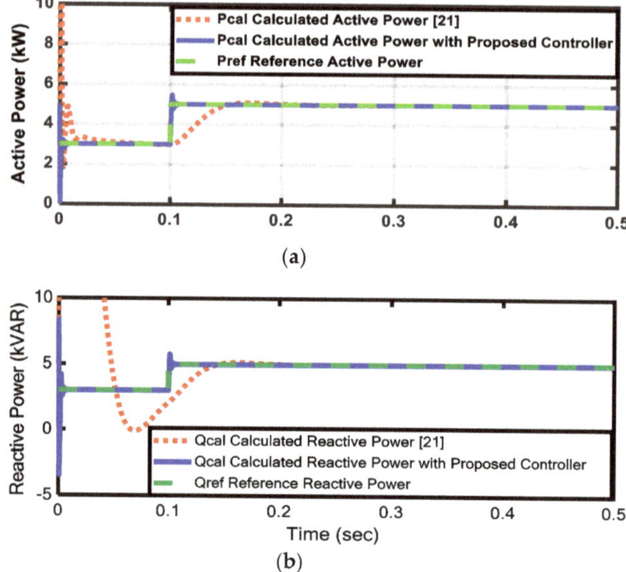

Figure 16. (**a**) Active power response with simultaneous step up change in real and reactive powers. (**b**) Reactive power response with simultaneous step up change in real and reactive powers.

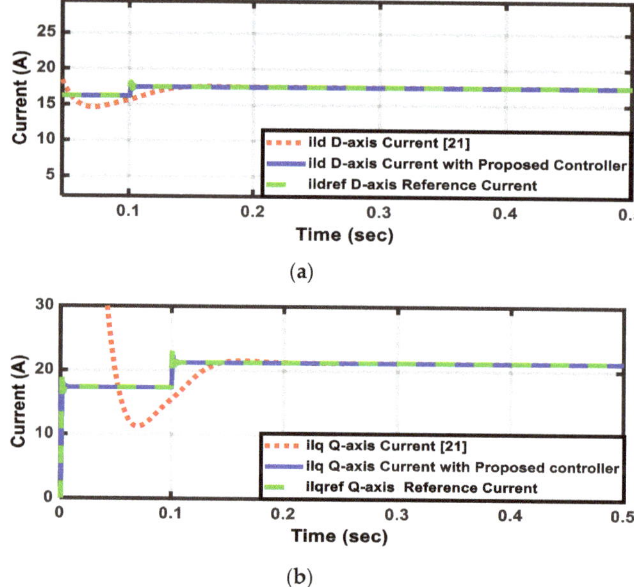

Figure 17. (**a**) D-axis current responses with simultaneous step change. (**b**) Q-axis current responses with simultaneous step change.

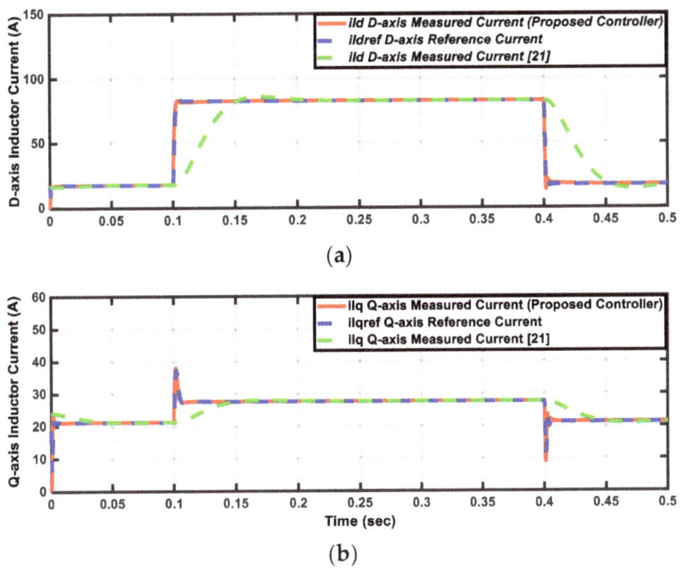

Figure 18. (**a**) D-axis current responses for two controllers at the fault disturbance. (**b**) Q-axis current responses for two controllers at the fault disturbance.

5.2. Two DGs Cases

In the second case, two different rated (5 KVA and 10 KVA) inverter-based DG units are controlled to bring their real and reactive powers with the grid as shown in Figure 2. Both units are tied together to share the injected powers to the grid. The proposed controller has been examined for different disturbances. Figure 19 shows the dynamic response of the calculated and reference real powers with different step changes for the two different DGs at different times. Firstly, the injected real powers of DG1 and DG2 have been stepped up from 7.5 kW to 10 kW at t = 0.1 sec and from 2.5 kW to 5 kW at t = 0.3 sec respectively. Figure 20 illustrates the dynamic response of the *dq* components of the inverter output currents related to this disturbance. Both injected real and reactive powers are mostly following the reference powers. It could be observed from the results that the controller has quick responses and track the references effectively. Secondly, Figure 21 depicts the dynamic response of the real and reactive powers with simultaneous step change in the injected real powers of DG1 and DG2. For DG1, the real power has been stepped down from 10 kW to 6 kW at t=0.1 sec then it has been stepped up from 6 kW to 10 kW at t = 0.4 sec. While for DG2, the real power has been stepped down from 5 kW to 2 kW at t = 0.3 sec then it has been stepped up from2 kW to 5 kW at t = 0.5 sec. With the proposed controllers, it is obvious that the response overshoot is not significant. The response of the microgrid equipped with the proposed controller keep better and fast reference tracking in a good way.

A severe disturbance has been applied to check the controller capability for such disturbance. At t = 0.1 sec, the microgrid lost DG1 and restore it again at t = 0.4 sec. Meanwhile, the real power of DG2 has been stepped down from 5 kW to 3 kW between t = 0.3 sec and t = 0.5 sec. Figure 22 shows the real and reactive power responses of DG1 and DG2 at this disturbance. While, Figure 23 shows the responses of the *dq* components of the inductor current of DG1 and DG2 at this disturbance. It can be concluded from this disturbance that the proposed controller keeps the microgrid operation in a better shape. Additionally, the proposed controller has a fast response tracking the reference and in a good way.

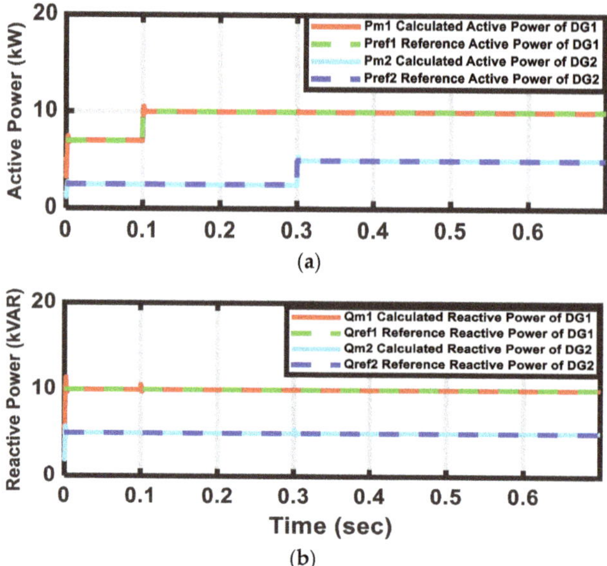

Figure 19. (**a**) Active power responses at stepped up active power disturbances at different times. (**b**) Reactive power responses at stepped up active power disturbances at different times.

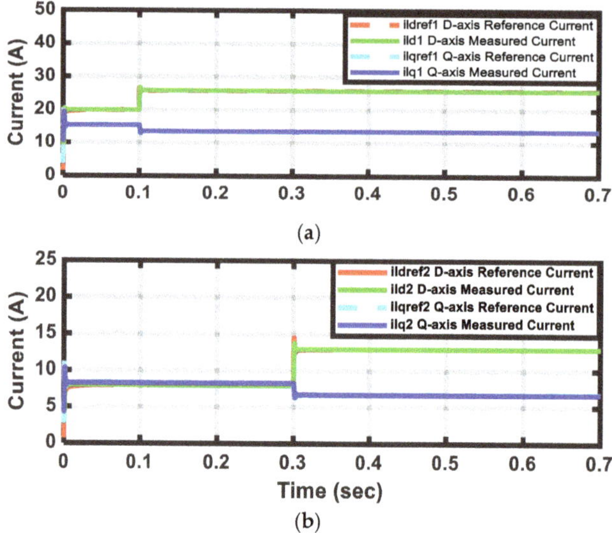

Figure 20. (**a**) D-axis current responses at stepped up active power disturbance. (**b**) Q-axis current responses at stepped up active power disturbance.

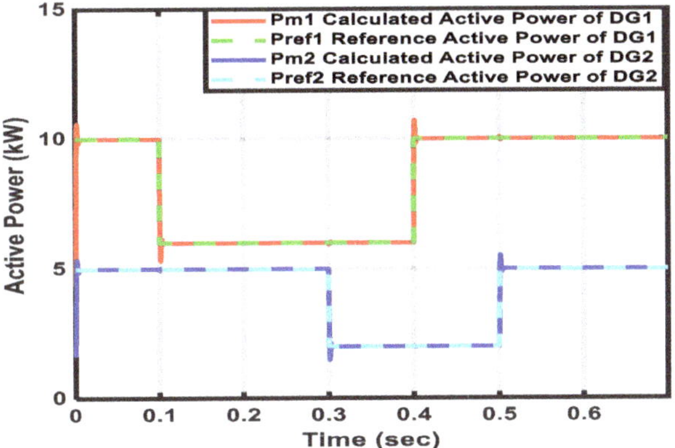

Figure 21. Power responses with stepped up and down real powers at different times.

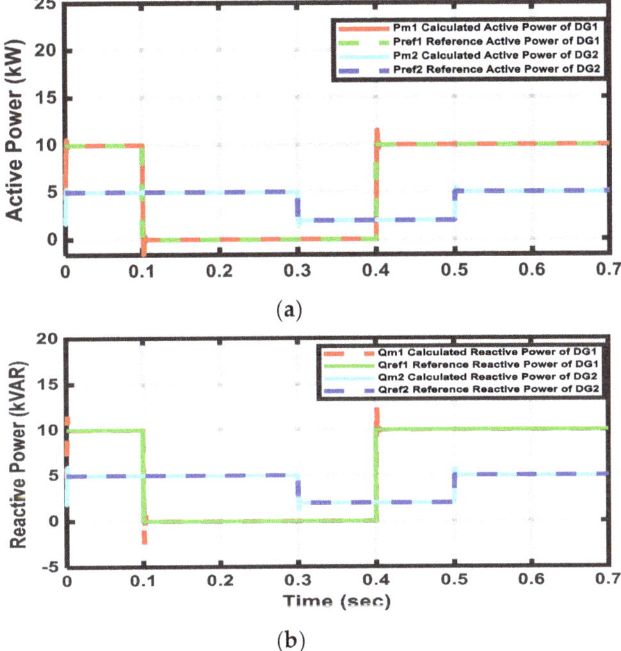

Figure 22. (**a**) Active power responses when the microgrid lost DG1 for 0.3 sec. (**b**) Reactive power responses when the microgrid lost DG1 for 0.3 sec.

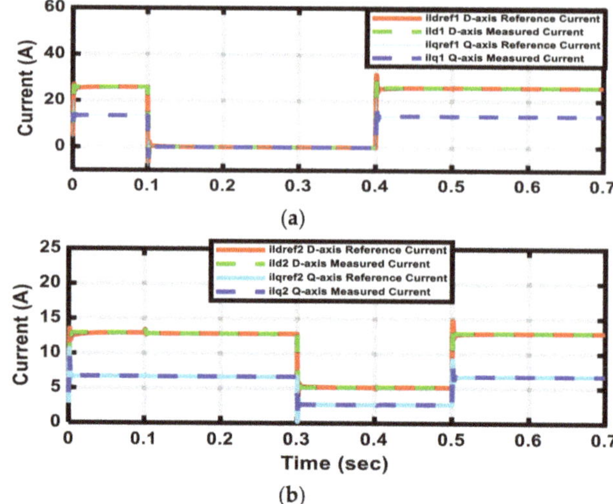

Figure 23. (**a**) D-axis current responses when the microgrid lost DG1 for 0.3 sec (**b**) Q-axis current responses when the microgrid lost DG1 for 0.3 sec.

6. Real-Time Implementation

The considered inverter-based DG with its proposed controller has been implemented on real-time digital simulator (RTDS) as shown in Figure 24 [34–36]. For verifying the theoretical simulation, the proposed microgrid is implemented and simulated in RTDS. RTDS works in real-time to provide solutions to power system equations quickly enough to accurately represent conditions in the real world [26]. RTDS offers superior accuracy over analogue systems. It allows for comprehensive product and/or configuration tests. RTDS provides a variety of transient study possibilities. With detailed models of power system components, a model resembles closely the real system setup in the RTDS. As RTDS works in continuous sustained real-time, the simulation is performed fast. With standard library blocks, using their physical representation, components such as grid, LC filter, inverter bridge and coupling inductor are modeled. The RTDS model of the grid-connected microgrid includes the models of the inverter, LC filter, the coupling impedance and the grid is given in Figure 24. The firing pulse generator and triangle wave generator blocks used for generating the firing pulses of the inverter gates are also included in Figure 24.

Additionally, the RTDS models of the power and current controllers are illustrated in Figure 25. The optimal controller parameters are engaged with the implemented RTDS model. Firstly, with stepping up the injected real and reactive powers from 3 kW to 5 kW and from 3 kVAR to 5 kVAR respectively, the dynamic response of the *dq* currents components, real and reactive powers are shown in Figures 26–29. Secondly, Figures 30 and 31 show the dynamic response of the real and reactive powers when a single-phase fault occurs at PCC. Finally, real and reactive powers have been stepped for the proposed system and for the work done previously in [21]. A comparison between the new and old controllers has been presented to prove the superiority of the proposed one. Figures 32 and 33 depict the dynamic response of the real and reactive powers for the previous and proposed controllers. As shown in the RTDS results, during the step changes, the controller has a reasonable capability to track the reference signal without significant overshoots. The proposed controller effectiveness under these disturbances is confirmed in RTDS. The given results illustrate the superiority of the proposed controller. The response of the microgrid equipped with the proposed controller keep better reference tracking in a good way. The results show the controller effectiveness under different disturbances.

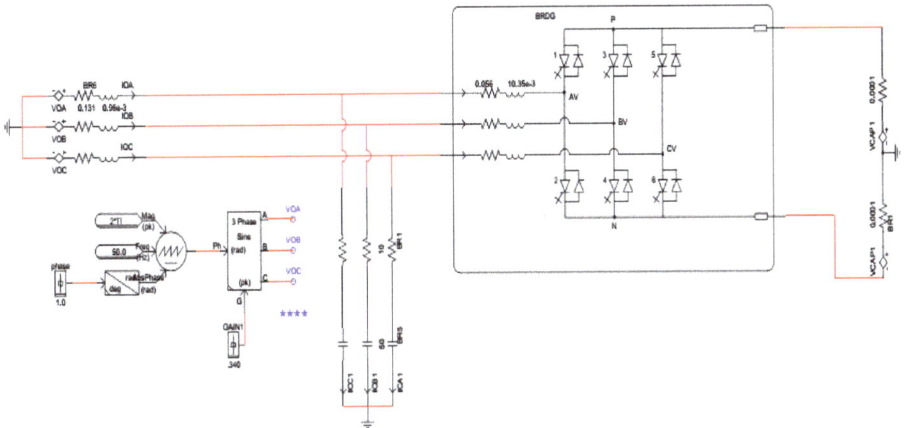

Figure 24. Laboratory setup for real-time digital simulation.

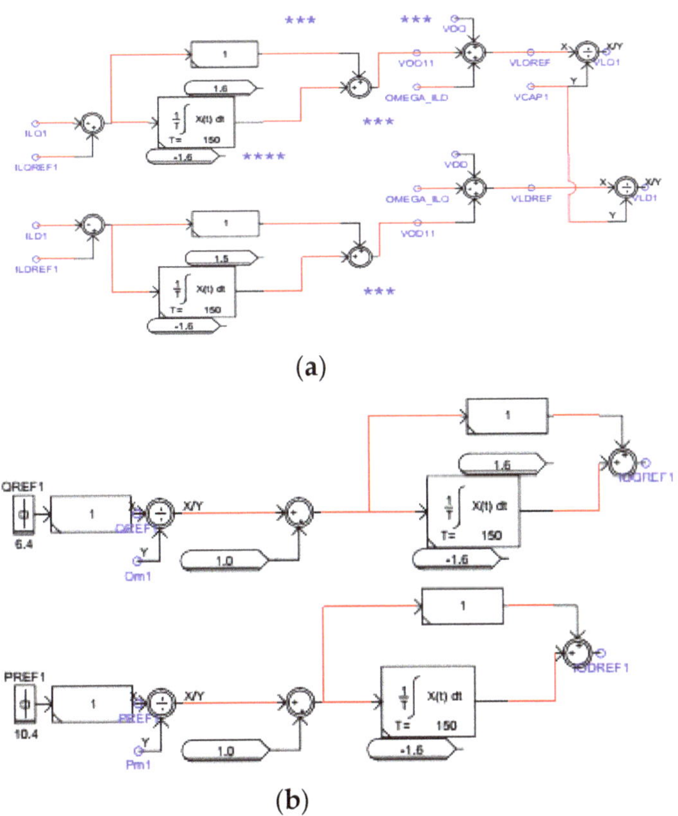

Figure 25. (**a**) Power controller on real-time digital simulation. (**b**) Current controller on real-time digital simulation.

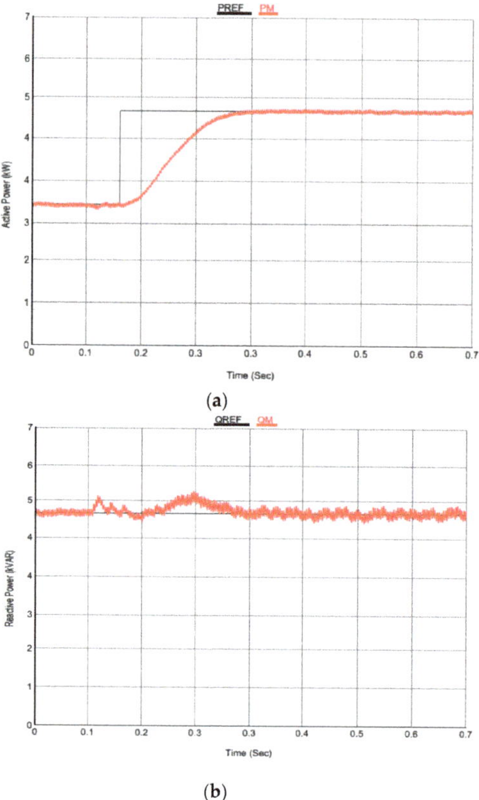

Figure 26. (a) Active power responses at stepped up real power disturbance. (b) Reactive power responses at stepped up real power disturbance.

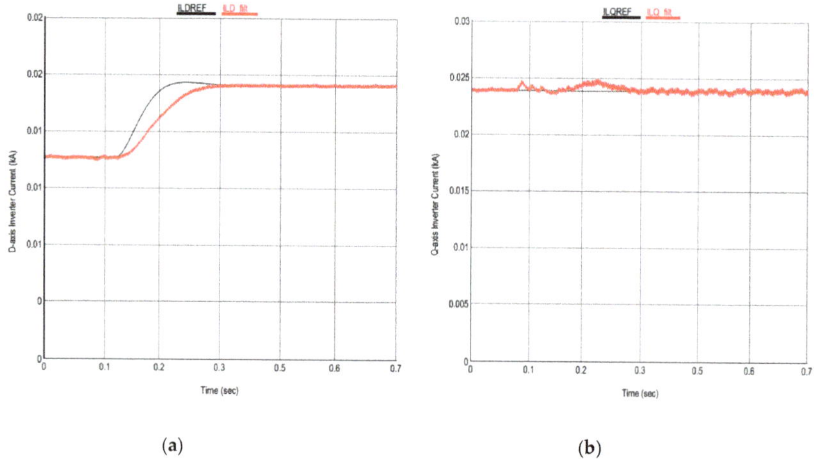

Figure 27. (a) D-axis inductor current response at stepped up real power disturbance. (b) Q-axis inductor current response at stepped up real power disturbance.

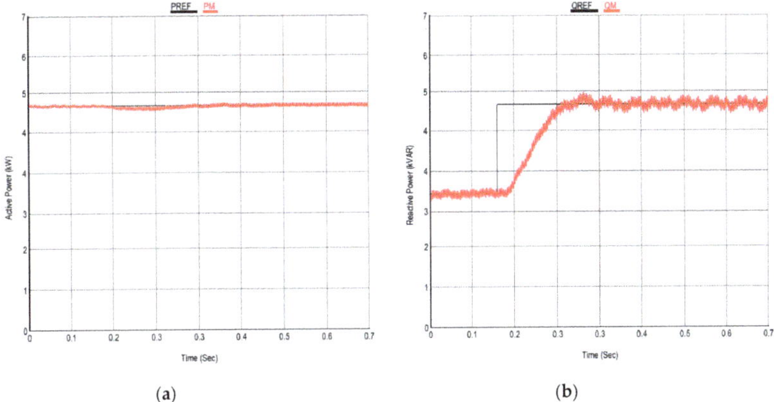

(a) (b)

Figure 28. (**a**) Active power responses at stepped up reactive power disturbance. (**b**) Reactive power responses at stepped up reactive power disturbance.

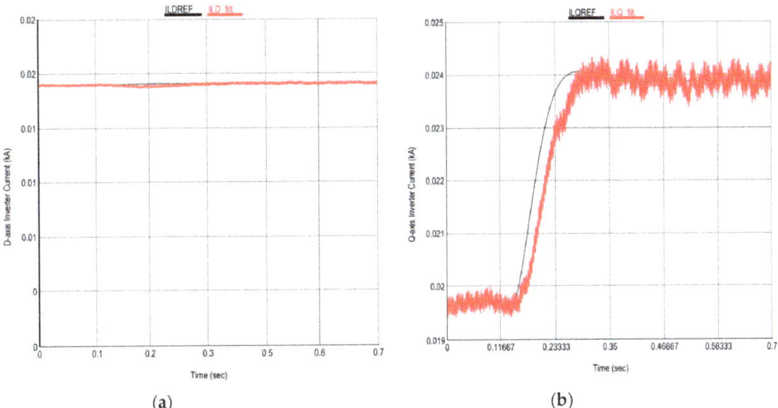

(a) (b)

Figure 29. (**a**) D-axis inductor current response at stepped up reactive power disturbance. (**b**) Q-axis inductor current response at stepped up reactive power disturbance.

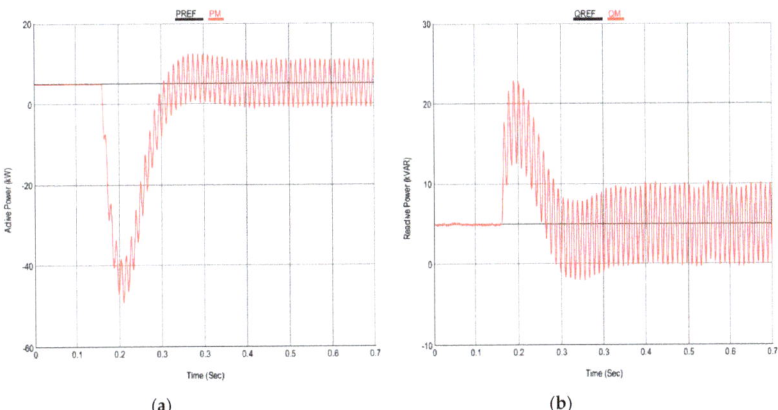

(a) (b)

Figure 30. (**a**) Active power responses at fault disturbance at PCC. (**b**) Reactive power responses at fault disturbance at PCC.

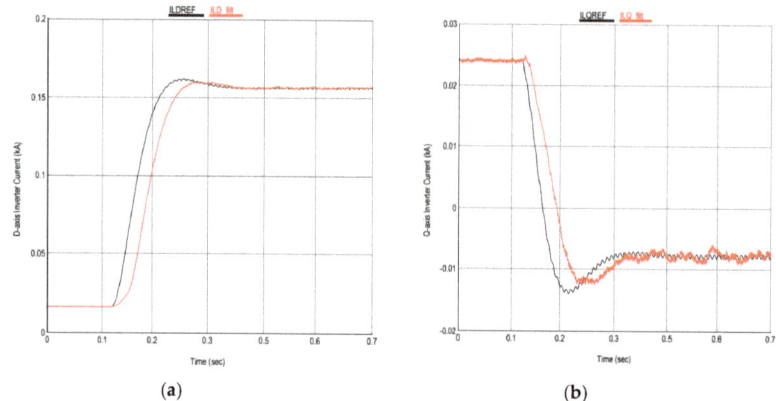

(**a**) (**b**)

Figure 31. (**a**) D-axis inductor current response at fault disturbance at PCC. (**b**) Q-axis inductor current response at fault disturbance at PCC.

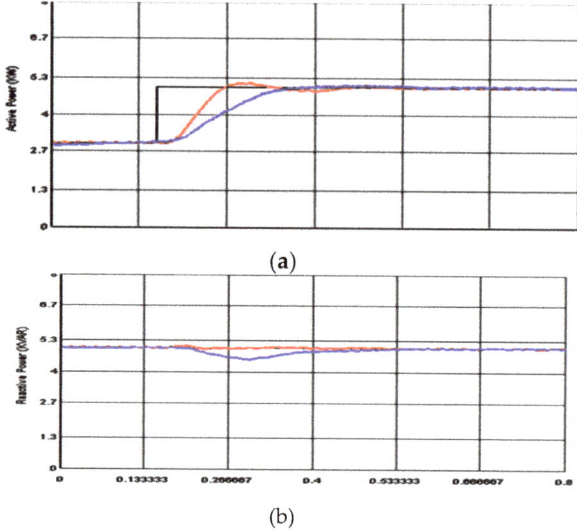

(**a**)

(b)

Figure 32. (**a**) Active power responses at stepped up active power disturbance. (**b**) Reactive power responses at stepped up active power disturbance.

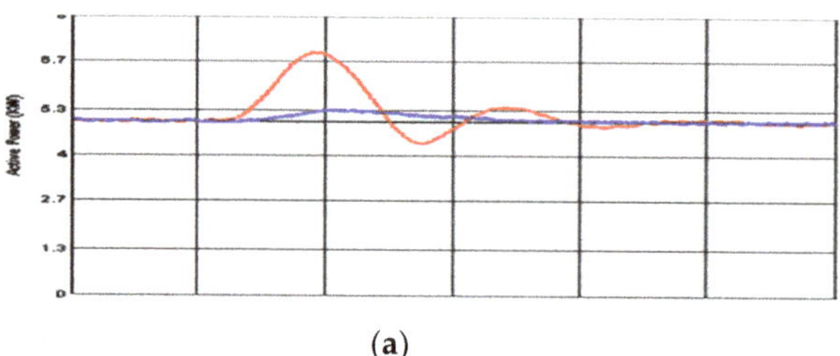

(**a**)

Figure 33. *Cont.*

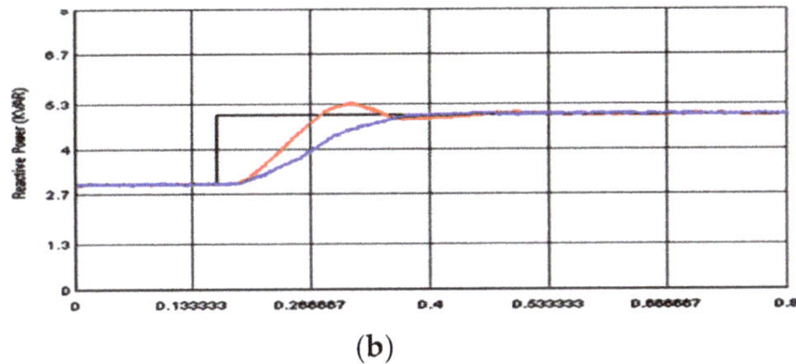

(b)

Figure 33. (**a**) Active power responses at stepped up reactive power disturbance. (**b**) Reactive power responses at stepped up reactive power disturbance.

7. Conclusion

In this paper, an optimal PQ control technique for inverter-based DG in the grid-connected microgrid has been presented. A power controller has been optimally designed to control a predefined injected real and reactive powers to the utility. Moreover, the current controller parameters and filter components have been optimized. To improve the microgrid dynamic response, the optimized controller parameters have been adjusted. For microgrid stability enhancement, with the proposed objective function, PSO has been employed for searching the optimal settings of the optimized controller and filter parameters. Two different cases in terms of the microgrid configuration have been considered to inject and share a predefined set real and reactive powers respectively. In both cases, the microgrid performance has been investigated for severe disturbances to clarify how the proposed optimal control improved the microgrid dynamic stability. The proposed controller response has been compared with that given in the literature. The time domain simulations confirm the proposed approach usefulness over the previous controller. For the proposed microgrid, the obtained RTDS results verify the usefulness of the proposed controllers for different disturbances. Additionally, RTDS results demonstrate reasonable performance with effective damping characteristics of the proposed controller.

Author Contributions: M.A.H. and M.A.A. initiated the idea, formulated the problem, performed the simulation, and analyzed the results. M.Y.W. participated in paper revision stage, contributed in enhancing the simulation results, and shared in paper writing.

Funding: This research was funded by King Fahd University of Petroleum and Minerals through the Power Research Group funded project RG171002 and King Abdullah City for Atomic and Renewable Energy (K.A. CARE).

Conflicts of Interest: The authors declare no conflict of interest.

References

1. Colmenar-Santos, A.; Reino-Rio, C.; Borge-Diez, D.; Collado-Fernández, E. Distributed generation: A review of factors that can contribute most to achieve a scenario of DG units embedded in the new distribution networks. *Renew. Sustain. Energy Rev.* **2016**, *59*, 1130–1148. [CrossRef]
2. Tavakoli, M.; Shokridehaki, F.; Akorede, M.F.; Marzband, M.; Vechiu, I.; Pouresmaeil, E. CVaR-based energy management scheme for optimal resilience and operational cost in commercial building microgrids. *Electr. Power Energy Syst.* **2018**, *100*, 1–9. [CrossRef]
3. Hassan, M.; Abido, M. Optimal design of microgrids in autonomous and grid-connected modes using particle swarm optimization. *IEEE Trans. Power Electron.* **2011**, *26*, 755–769. [CrossRef]
4. Mohammadi, F. Power Management Strategy in Multi-Terminal VSC-HVDC System. In Proceedings of the 4th National Conference on Applied Research in Electrical, Mechanical Computer and IT Engineering, Tehran, Iran, 4 October 2018.
5. Mohammadi, F.; Nazri, G.A.; Saif, M. A Bidirectional Power Charging Control Strategy for Plug-in Hybrid Electric Vehicles. *Sustainability* **2019**, *11*, 4317. [CrossRef]

6. Hassan, M.A.; Worku, M.Y.; Abido, M.A. Optimal design and real time implementation of autonomous microgrid including active load. *Energies* **2018**, *11*, 1109. [CrossRef]
7. Danish, M.S.S.; Matayoshi, H.; Howlader, H.O.R.; Chakraborty, S.; Mandal, P.; Senjyu, T. Microgrid Planning and Design: Resilience to Sustainability. In Proceedings of the 2019 IEEE PES GTD Grand International Conference and Exposition Asia (GTD Asia), Bangkok, Thailand, 19–23 March 2019; pp. 253–258.
8. Danish, M.S.S.; Senjyu, T.; Funabashi, T.; Ahmadi, M.; Ibrahimi, A.M.; Ohta, R.; Howlader, H.O.R.; Zaheb, H.; Sabory, N.R.; Sediqi, M.M. A sustainable microgrid: A sustainability and management-oriented approach. In *Proceedings of the Applied Energy Symposium and Forum, Renewable Energy Integration with Mini/Microgrids*; Energy Procedia: Rhodes, Greece, 2018; Volume 159, pp. 160–167.
9. Sahoo, S.K.; Sinha, A.K.; Kishore, N.K. Control techniques in AC, DC, and hybrid AC–DC microgrid: A Review. *IEEE J. Emerg. Sel. Top. Power Electron.* **2018**, *6*, 738–759. [CrossRef]
10. Xin, H.; Zhang, L.; Wang, Z.; Gan, D.; Wong, K.P. Control of island AC microgrids Using a fully distributed approach. *IEEE Trans. Smart Grid* **2015**, *6*, 943–945. [CrossRef]
11. Bai, W.; Lee, K. Distributed Generation System Control Strategies in Microgrid Operation. Preprints of the 19th World Congress. The International Federation of Automatic Control, Cape Town, South Africa, 24–29 August 2014; pp. 11938–11943.
12. Serban, I.; Marinescu, C. Control strategy of three-phase battery energy storage systems for frequency support in microgrids and with uninterrupted supply of local loads. *IEEE Trans. Power Electron.* **2014**, *29*, 5010–5020. [CrossRef]
13. Jena, S.; Babu, C.; Mishra, G.; Naik, A. Reactive power compensation in inverter-interfaced distributed generation. In Proceedings of the 2011 International Conference on Energy, Automation, and Signal (ICEAS), Bhubaneswar, India, 28–30 December 2011; pp. 1–6.
14. Hornik, T.; Zhong, Q. A Current-control strategy for voltage-source inverters in microgrids based on H∞ and repetitive control. *IEEE Trans. Power Electron.* **2011**, *26*, 943–952. [CrossRef]
15. Zhang, N.; Tang, H.; Yao, C. A systematic method for designing a PR controller and active damping of the LCL filter for single-phase grid-connected PV inverters. *Energies* **2014**, *7*, 3934–3954. [CrossRef]
16. Jeong, H.; Kim, G.; Lee, K. Second-order harmonic reduction technique for photovoltaic power conditioning systems using a proportional-resonant controller. *Energies* **2013**, *6*, 79–96. [CrossRef]
17. Yao, Z.; Xiao, L.; Guerrero, J. Improved control strategy for the three-phase grid-connected inverter. *IET Renew. Power Gener.* **2015**, *9*, 587–592. [CrossRef]
18. Kumar, V.; Jees, S.; Gomathi, V. Control techniques for single phase inverter to interface renewable energy sources with the microgrid. *Int. J. Adv. Res. Electr. Electron. Instrum. Eng.* **2014**, *3*, 437–447.
19. Lia, S.; Jaithwaa, I.; Suftaha, R.; Fua, X. Direct-current vector control of three-phase grid-connected converter with L, LC, and LCL filters. *Electr. Power Compon. Syst.* **2015**, *43*, 1644–1655. [CrossRef]
20. Schonardie, M.; Coelho, R.; Schweitzer, R.; Martins, D. Control of the active and reactive power using dq0 transformation in a three-phase grid-connected PV system. In Proceedings of the 2012 IEEE International Symposium Industrial Electronic, Hangzhou, China, 28–31 May 2012; pp. 264–269.
21. Hassan, M.; Abido, M. RTDS implementation of the optimal design of the grid-connected microgrids using particle swarm optimization. In Proceedings of the International Conference on Renewable Energies and Power Quality (ICREPQ'12), Santiago de Compostela, Spain, 28–30 March 2012.
22. Dai, M.; Marwali, M.N.; Keyhani, A. Power flow control of a single distributed generation unit. *IEEE Trans. Power Electron.* **2008**, *23*, 343–352. [CrossRef]
23. Jang, M.; Mihai, C.; Agelidis, V.G. A single-phase grid-connected fuel cell system based on a boost-inverter. *IEEE Trans. Power Electron.* **2013**, *28*, 279–288. [CrossRef]
24. Fazeli, S.M.; Ping, H.W.; Rahim, N.B.A.; Ooi, B.T. Individual-phase decoupled PQ control of three-phase voltage source converter. *IET Gener. Transm. Distrib.* **2013**, *7*, 219–1228. [CrossRef]
25. Adhikari, S.; Li, F.X.; Li, H.J. PQ and PV control of photovoltaic generators in distribution systems. *IEEE Trans. Smart* **2015**, *6*, 2929–2941. [CrossRef]
26. Chen, M.; Wang, H.; Zeng, G.; Dai, Y.; Bi, D. Optimal P-Q control of grid-connected inverters in a microgrid based on adaptive population extremal optimization. *Energies* **2018**, *11*, 2107. [CrossRef]
27. Hassan, M.; Abido, M. Real time implementation and optimal design of autonomous microgrids. *Electr. Power Syst. Res.* **2014**, *109*, 118–127. [CrossRef]

28. Patel, R.; Li, C.; Meegahapola, L.; McGrath, B.; Yu, X. Enhancing optimal automatic generation control in a multi-area power system with diverse energy resources. *IEEE Trans. Power Syst.* **2019**, *34*, 3465–3475. [CrossRef]
29. Hassan, M.; Abido, M. Optimal power sharing of an inverter–based autonomous microgrid. In Proceedings of the Conference on Renewable Energies and Power Quality (ICREPQ'13), Bilbao, Spain, 20–22 March 2013.
30. Al-Saedi, W.; Lachowicz, S.W.; Habibi, D.; Bass, O. Voltage and frequency regulation-based DG unit in an autonomous microgrid operation using particle swarm optimization. *Int. J. Electr. Power Energy Syst.* **2013**, *53*, 742–751. [CrossRef]
31. Dong, D.; Wen, B.; Boroyevich, D.; Mattavelli, P.; Xue, Y. Analysis of phase-locked loop low-frequency stability in three-phase grid-connected power converters considering impedance interactions. *IEEE Trans. Ind. Electron.* **2015**, *62*, 310–321. [CrossRef]
32. Errami, Y.; Ouassaid, M.; Maaroufi, M. Modelling and optimal power control for permanent magnet synchronous generator wind turbine system connected to utility grid with fault conditions. *World J. Model. Simul.* **2015**, *11*, 123–135.
33. Kennedy, J.; Eberhart, R. Particle swarm optimization. In Proceedings of the IEEE International Conference on Neural Networks, Perth, Australia, 27 November–1 December 1995; pp. 1942–1948.
34. Worku, M.; Hassan, M.; Abido, M. Real time energy management and control of renewable energy based microgrid in grid connected and island modes. *Energies* **2019**, *12*, 276. [CrossRef]
35. RTDS Technologies. *Real Time Digital Simulator Power System and Control User Manual*; RTDS Technologies: Winnipeg, MB, Canada, 2009.
36. Mahmoud, M.; Sattar, A. Real time implementation of the grid to analyze the performance of the variable speed wind turbine–generator during grid disturbances. *Int. J. Comput. Electr. Eng.* **2016**, *8*, 104–116. [CrossRef]

Article

An Improved Power Management Strategy for MAS-Based Distributed Control of DC Microgrid under Communication Network Problems

Thanh Van Nguyen and Kyeong-Hwa Kim *

Department of Electrical and Information Engineering, Seoul National University of Science and Technology, 232 Gongneung-ro, Nowon-gu, Seoul 01811, Korea; vanthanhpro94@gmail.com
* Correspondence: k2h1@seoultech.ac.kr; Tel.: +82-2-970-6406

Received: 29 November 2019; Accepted: 20 December 2019; Published: 22 December 2019

Abstract: In this paper, an improved power management strategy (PMS) for multi-agent system (MAS)-based distributed control of DC microgrid (DCMG) under communication network problems is presented in order to enhance the reliability of DCMG and to ensure the system power balance under various conditions. To implement MAS-based distributed control, a communication network is constructed to exchange information among agents. Based on the information obtained from communication and local measurements, the decision for the local controller and communication is optimally given to guarantee the system power balance under various conditions. The operating modes of the agents can be determined locally without introducing any central controller. Simultaneously, the agents can operate in a deliberative and cooperative manner to ensure global optimization by means of the communication network. Furthermore, to prevent the system power imbalance caused by the delay in grid fault detection and communication in case of the grid fault, a DC-link voltage (DCV) restoration algorithm is proposed in this study. In addition, to avoid the conflict in the DCV control among power agents in case of the grid recovery under communication failure, a grid recovery identification algorithm is also proposed to improve the reliability of DCMG operation. In this scheme, a special current pattern is generated on the DC-link at the instant of the grid recovery by the grid agent, and other power agents identify the grid recovery by detecting this current pattern. Comprehensive simulations and experiments based on DCMG testbed have been carried out to prove the effectiveness of the PMS and the proposed control schemes under various conditions.

Keywords: communication network problems; DC microgrid; distributed control; improved power management; multi-agent systems; grid recovery

1. Introduction

In recent years, the concept of microgrids (MGs) has been introduced as an effective and potential solution to integrate various renewable energy sources (RESs) such as wind and solar into the grid [1]. For the purpose of stabilizing the system operation under the intermittent nature of RESs and continuous variations of load demand, energy storage systems (ESSs) are usually used in MGs [2]. With the development of technology, electric vehicles can be utilized as multiple ESSs to stabilize the system operation of MGs by precise control strategy and simultaneously to regulate the voltage and frequency of the power grid [3]. Under the normal grid conditions, MG operates in the grid-connected mode in which the system power balance is ensured by the grid. In the case of grid fault, however, the MG is completely independent of the grid, operating in the islanded mode. In this sense, the system power balance of MG should be achieved by means of a coordinated operation of power units such as RESs, ESSs, and grid according to load demands [4]. As the reliability issue becomes a primary interest, there has been an approach to develop a fast proactive hybrid DC circuit breaker which interrupts the DC fault [5]. As another approach to enhance the reliability, the DC-link voltage (DCV) control method has been presented for MG by the conventional centralized power flow control strategy [6].

Depending on the types of bus voltage, MGs can be mainly classified into DC microgrids (DCMGs), AC microgrids (ACMGs), and hybrid AC/DC microgrids [7]. Among them, DCMGs are known to be more attractive due to several advantages as compared with the other configurations. DCMGs provide a convenient interface in connecting various DC loads and operate with better efficiency in transmission and distribution. Moreover, the consideration for the harmonic injection and frequency stability is not necessary. As a result, DCMGs can be considered as an efficient, reliable, and cost-effective option in some applications [8].

In view of the communication perspective, the control of DCMGs can be divided into three methods: centralized control, decentralized control, and distributed control [9]. In centralized control, the data from distributed power units are collected in a central controller (CC). Then, the CC processes the acquired data to send the feedback commands back to units via digital communication links (DCLs). The centralized control normally suffers from many drawbacks related to the single point of failure, reliability, flexibility, and scalability [10]. On the other hand, the decentralized approach can provide high reliability and flexibility due to the absence of CC and DCLs among distributed power units in the system [11]. Nevertheless, it is quite difficult to ensure the global optimization of the entire system because each distributed unit lacks the information of others. By analyzing two above control methods, the distributed control is introduced as an optimal solution which combines the advantages as well as restricts the disadvantages of both methods [12]. In particular, the single point of failure as in the centralized control can be avoided since CC does not exist in the system. In addition, as compared to the decentralized approach, the problem of global optimization is effectively solved by implementing the DCLs among neighboring units. Therefore, a distributed control is considered as a potential alternative for the control of the power system in the future.

For the distributed control, a multi-agent system (MAS) is employed as an advanced technique which is able to handle complex problems in DCMGs including the system power balance, reliability, and stability in a more flexible and intelligent way [13]. Therefore, several studies relied on the MAS-based distributed control strategy of DCMG have been presented recently in the literature [14–18]. In [14], an intelligent control based on the MAS technique is presented to control the MG in both the grid-connected and islanded modes. In this scheme, the optimization in real-time control of the MG is achieved by the negotiation among agents to share the available energy. In [15], a novel distributed optimal control which applies the consensus algorithm-based MAS is introduced in order to realize the optimal control as well as to provide fast frequency recovery under various load conditions. With the aim of developing secondary control by MAS-based distributed control for the frequency recovery in islanded MG, a distributed robust control strategy is presented in [16]. To obtain the accurate reactive power-sharing among agents, a MAS-based hierarchical distributed coordinated control strategy is proposed in [17]. In [18], both reactive power-sharing and voltage control are taken into account by implementing a distributed MAS-based finite-time consensus algorithm in two layers. Although the aforementioned schemes can achieve key functions of control for MG such as the system stability, energy management, active and reactive power controls, the implementation of system control only focuses on purely inverter-based distribution generations (DGs). However, in practice, most present MG applications involve the networks of mixed agents such as inverter-based DGs, RESs, ESSs, and loads. Because these MG components have different characteristics, each type of agent should be designed considering a specific control strategy during operation. Furthermore, the problems of communication network in the implementation of MAS-based distributed control have not been considered yet.

In the MAS-based distributed control, the communication network problems such as delay or failure are ubiquitous during the process of information transmission among agents, which may cause the system malfunction, instability, or even collapse [19]. In order to analyze the effect of communication network problems on the system stability, the control scheme has been tested under the presence of time delay in [20,21]. However, since the effect of time delay on system performance has been only validated by the simulation tests in these studies, the reliability of system operation is not completely ensured in reality. Moreover, as shown in the simulation results of these studies, larger time delay results in longer and stronger fluctuation of the controlled variables such as the voltage and frequency in case of load variation. For the purpose of improving the control performance further, the time delay is considered in the designs of the secondary controllers [22,23]. In [24,25], a local controller based on the delay-dependent H_∞ method is presented to deal with the transmission time delays. Although the schemes in [22–25] can guarantee the system stabilization, the control performance is still limited due to the requirement of an allowable maximum upper bound for the time delay. Motivated by this concern, a communication-based control strategy that can maintain the system stability with unbounded time delays is presented in [26]. In comparison to time delay, the communication failure is known to be more serious in the communication-based systems. To address the drawback of the data loss caused by communication failure, a prediction scheme is constructed in [27], in which each agent forecasts the lost data by using an extreme learning machine.

As mentioned earlier, the problems of the communication network result in the performance deterioration or even system instability, which is more serious in cases of grid fault and grid recovery. When DCMG operates in the grid-connected mode, the system power balance is normally achieved by implementing the DC-link voltage control mode (DCVM) by the grid agent. When the grid fault occurs, the grid agent informs the other agents of the fault state via the communication links, and then, the DCMG operation is switched into the islanded mode. After receiving the grid fault information, the remaining agents automatically undertake the system power balance by implementing their own DCVM. Unfortunately, however, unexpected communication problems prevent the remaining agents from recognizing the grid fault information instantly. As a result, any agents cannot serve to achieve the system power balance. On the contrary, as soon as the grid is recovered from the fault, the grid agent should inform the recovery state to the other agents, and then switch the operation mode into DCVM. In compliance with this mode change, the remaining agents should stop their DCVM operation, turning over the authority of DCV to the grid agent. However, the remaining agents may fail to recognize the grid recovery in the presence of communication problems, which causes the conflict in the system control because two voltage control sources exist in DCMG.

To deal with the aforementioned drawbacks caused by the communication network problems in both the grid fault and grid recovery cases, an improved power management strategy (PMS) using MAS-based distributed control of DCMG is presented in this paper. In this study, DCMG consists of a grid agent, a battery agent, a wind power generation system (WPGS) agent, and a load agent. To ensure the system power balance under various conditions, each agent investigates the information obtained from both the local measurement and the neighboring agents via the communication lines. Then, the decision for local control mode and communication data is optimally made for the system power balance. By using this control scheme, the control mode of agents can be determined locally without any intervention of CC, which effectively avoids the single point of failure as in the centralized control. Also, all the agents can operate in a deliberative and cooperative manner to ensure globally optimal operation by means of the communication network. In addition, to deal with the impact of communication problems in the case of the grid fault, a DCV restoration algorithm is introduced to restore the DCV stably to its nominal value. Furthermore, to recognize the grid recovery reliably in other agents even under the communication failure, the grid recovery identification algorithm is introduced. For this purpose, a special current pattern is generated on DC-link by the grid agent once the grid is recovered. By detecting this current pattern on DC-link, the remaining agents can reliably identify the grid recovery even without the communication. To validate the feasibility of the MAS-based distributed control as well as the proposed schemes under the communication problems, the simulations based on the PSIM software and the experiments based on laboratory prototype DCMG testbed are carried out.

This paper is organized as follows: Section 2 describes the system configuration of DCMG with MAS-based distributed control. Power management and control strategies of local agents are discussed in detail in Section 3. Section 4 presents the proposed control strategies under communication network problems. The simulation and experimental results are given in Sections 5 and 6, respectively. Finally, Section 7 concludes the paper.

2. System Configuration of DCMG with MAS-based Distributed Control

2.1. MAS-based Distributed Control of DCMG

Figure 1 shows the configuration of DCMG with the MAS-based distributed control approach, in which system units are represented by corresponding autonomous agents, namely, grid agent, battery agent, WPGS agent, and load agent. The functions of each agent are summarized as follows.

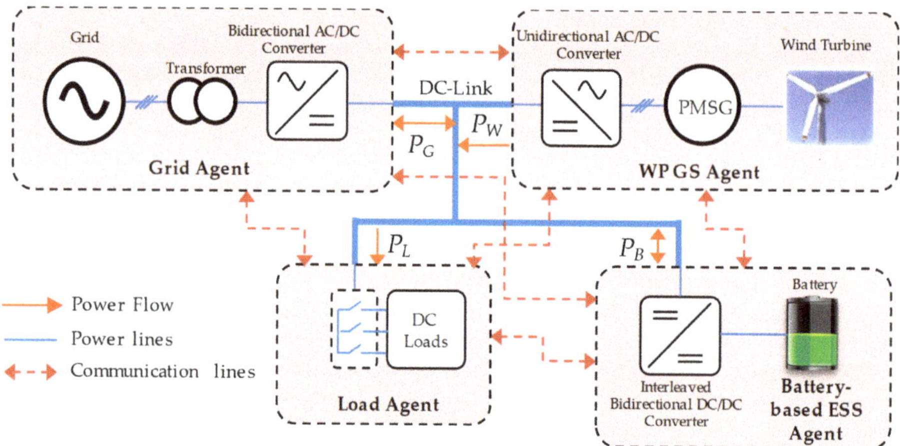

Figure 1. DCMG with MAS-based distributed control.

Grid agent: the grid agent receives the information on the grid statuses such as the normal, fault, or recovery from the grid operator (GO) to determine the operation mode of DCMG in the grid-connected or islanded. In the grid-connected mode, the grid agent ensures the supply–demand power balance in DCMG by controlling the exchange power between DCMG and grid within the maximum exchange power. In addition, the grid agent is responsible for the seamless transfer between the grid-connected and islanded modes.

Battery agent: the battery agent receives the information on the state of DCV control from the grid agent via a communication line. If the grid agent is incapable of controlling the DCV, the battery agent is switched into the DCVM to regulate the DCV at the nominal value. Furthermore, by collecting the state of charge (*SOC*), battery voltage, and battery current, the battery agent realizes relevant control modes to ensure that the battery is operated in a safe range.

WPGS agent: generally, the WPGS agent is operated in order to extract the maximum power from the wind into DC-link by the maximum power point tracking (MPPT) mode. However, when both the grid and battery agents are incapable of regulating the DCV, the WPGS agent changes the operation from the MPPT mode to DCVM to regulate the DCV.

Load agent: the load agent is responsible for monitoring the load demand, and also, providing the information on load demand to other agents. Another role of this agent is to implement the load shedding (SHED) and load reconnection (RECO) to keep the system stable as well as to optimize the available power on DC-link.

As can be seen in Figure 1, each agent can not only monitor and control the corresponding power units but also communicate with other agents through the communication network. By using the communication among agents, the system control is realized in a distributed way in which each agent makes its control decision locally. This enhances the control performance and system reliability as compared with the centralized and decentralized control approaches.

2.2. System Communication Topology

In general, the communication network plays an important role in MAS-based distributed control. Figure 2 presents the system communication topology used in this study, in which each agent communicates with others via communication lines. Defining an appropriate communication network that specifies the data information exchanged among agents is the crucial step in the design procedure of MAS-based distributed control [28]. In this paper, the exchange data, data type, and data information are described in Table 1 in detail. For example, the maximum supplied power, the maximum injected power, and the grid states are transmitted through the communication line 1 into the grid agent in the format of double and binary data.

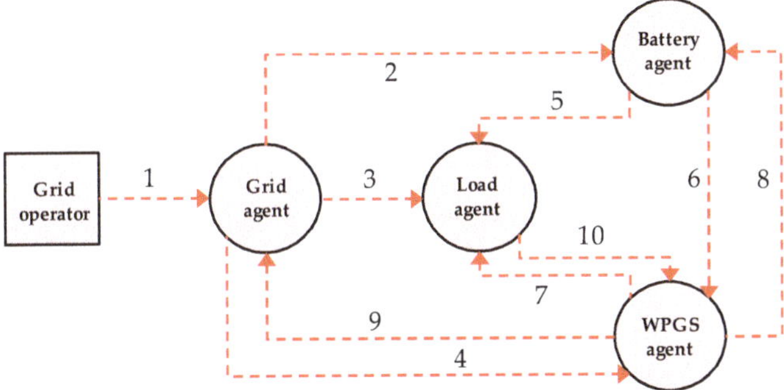

Figure 2. System communication topology.

Table 1. Detailed description of exchange data.

Communication Line	Exchange Data	Data Type	Data Information
1	$P^{max}_{G,rec}$	Double	Maximum power supplied from grid to DCMG
	$P^{max}_{G,inv}$	Double	Maximum power injected to grid from DCMG
	G^{state}	Binary	00: Grid has fault.
			11: Grid is recovered.
			01: Grid is being normal.
2, 3, 4	G^{ctrl}	Binary	1. Grid agent is able to control DCV.
			0: Grid agent is not able to control DCV.
5, 6	B^{ctrl}	Binary	1: Battery agent is able to control DCV.
			0: Battery agent is not able to control DCV.
7	W^{ctrl}	Binary	1: WPGS agent is able to control DCV.
			0: WPGS agent is not able to control DCV.
8, 9	P_{W-L}	Binary	1: Load power is greater than wind power ($P_L \geq P_W$).
			0: Load power is smaller than wind power ($P_L < P_W$).
10	P_L	Double	Load power

3. Power Management and Control Strategy of Local Agents

3.1. Control Strategy of Grid Agent

Figure 3 shows the control strategy which is designed in this study for local operation of the grid agent in MAS-based distributed control approach. As can be seen, the entire algorithm is divided into three layers, namely, the information collection, the decision of control mode, and the decision of communication. In the information collection layer, the information is gathered from the local measurement as well as GO and neighbor agents. The GO provides the information on the maximum exchange powers and grid states to the grid agent through the communication line 1 as in Table 1. Local measurement and neighbor agents offer the information on the voltage, current, and supply–demand power relationship. After gathering the necessary information, the grid agent determines appropriate control modes as shown in the decision of control mode layer. If the grid has a fault, the grid agent switches the operation to IDLE mode, and the variable G^{ctrl} is set to 0, which indicates that the grid agent is incapable of controlling the DCV. This information is transmitted to other agents.

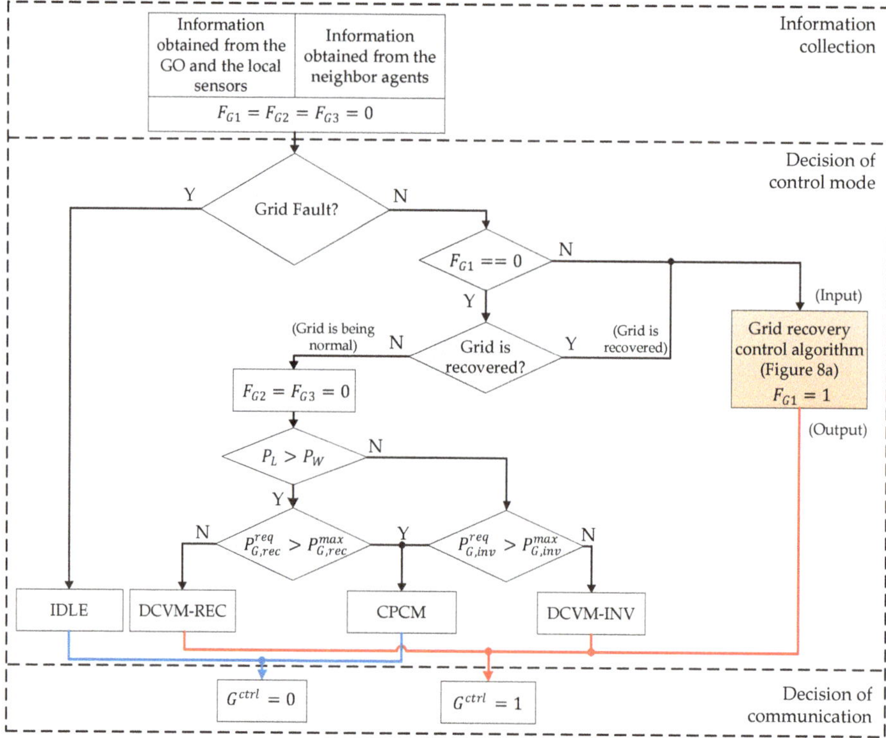

Figure 3. Control strategy of grid agent.

When the grid fault does not occur, the grid agent further investigates whether the grid is being normal or the grid is recovered. If the grid is being normal, the grid agent determines the operation as DCVM-REC or DCVM-INV to regulate the DCV and to guarantee the system power balance depending on the supply–demand power relationship. While the operation DCVM-REC compensates the power deficit by injecting the power from the grid to DCMG, the operation DCVM-INV makes the grid absorb the power surplus on DC-link. When the required power to maintain the system power balance becomes larger than the maximum level obtained from GO, the grid agent switches the operation into constant power control mode (CPCM). Since the exchange power between the grid and DCMG is at the maximum possible level under the CPCM, the grid agent is not able to control the DCV.

In case that the grid is recovered from the fault, a grid recovery control algorithm is triggered. The details of this algorithm and function for the flags F_{G1}, F_{G2}, and F_{G3} will be explained in Section 4. After the decision of local control mode is released, the information on the ability of the DCV control by the grid agent is transmitted to other agents via communication lines.

Based on the control strategy of the grid agent in Figure 3, the local control block is implemented to realize all the operating modes [6].

3.2. Control Strategy of Battery Agent

Figure 4 shows the control strategy which is designed in this study for local operation of the battery agent in MAS-based distributed control approach. Similarly, the entire algorithm is divided into three layers. In the information collection layer, the information is collected from the local measurement and neighbor agents. By using the local measurement, the information of battery state, *SOC*, voltage, and current can be obtained locally. The information on the supply–demand power relationship and the ability of the DCV control by the grid agent is obtained via communication lines from neighbor agents. In this figure, the DCV restoration, the grid recovery identification algorithm, and the function of flags F_{B1} and F_{B2} will be explained in Sections 4.1 and 4.2.

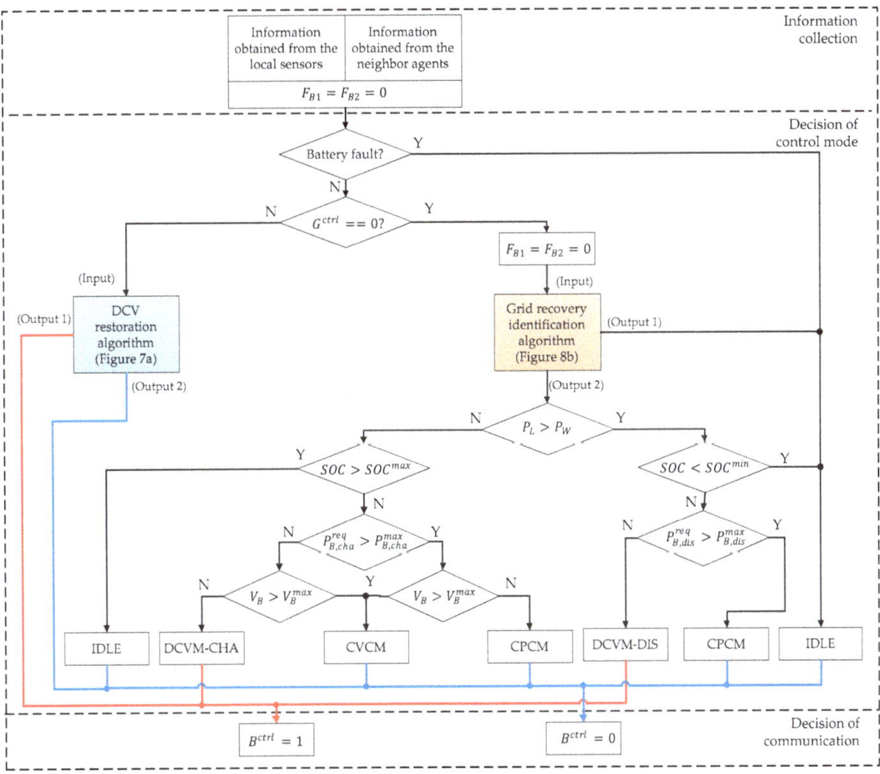

Figure 4. Control strategy of battery agent.

If the battery has a fault, the battery agent switches the operation into IDLE mode and the variable B^{ctrl} is set to 0 to imply that the battery agent is incapable of controlling the DCV. This information is also transmitted to other agents.

In case the battery is normal, the battery agent determines appropriate control modes according to the supply–demand power relationship, battery SOC, voltage, and current to ensure the system power balance in DCMG under various conditions. When the battery statuses such as SOC, voltage, and required power are within their predefined ranges, the battery agent determines the operation as DCVM-DIS to inject the power to DC-link or DCVM-CHA to absorb the power from DC-link for the purpose of regulating the DCV to its nominal value. If the battery SOC goes beyond the safe range defined by SOC^{min} and SOC^{max}, the battery agent switches the operation into IDLE mode to avoid overcharging or overdischarging. In addition, if the required charging power of $P_{B,cha}^{req}$ and the required discharging power of $P_{B,dis}^{req}$ exceed the maximum levels of $P_{B,cha}^{max}$ and $P_{B,dis}^{max}$, the operation CPCM is realized to charge/discharge the battery with its maximum capability. As a result, the overheat or damage during the battery operation can be avoided. Similarly, in order to prevent the battery from overcharging voltage, the constant voltage control mode (CVCM) is realized to charge the battery with the maximum voltage level. After the decision of local control mode is given, the information on the ability of the DCV control by the battery agent (B^{ctrl}) is simultaneously informed to other agents via communication lines.

Based on the control strategy of the battery agent in Figure 4, the local control block is implemented to realize all the operating modes [6].

3.3. Control Strategy of WPGS Agent

Figure 5 shows the control strategy for local operation of the WPGS agent in MAS-based distributed control approach. Similar to the above agents, the entire algorithm is also divided into three layers. In the information collection layer, the information is obtained from the local measurement and neighbor agents. By using the local measurement, the extracted power from wind turbine (P_W) can be obtained. The information on the supply–demand power relationship and the ability of the DCV control of battery and grid agent is gathered from neighbor agents. In this figure, the DCV restoration, the grid recovery identification algorithm, and the function of flag F_W will be presented in Sections 4.1 and 4.2.

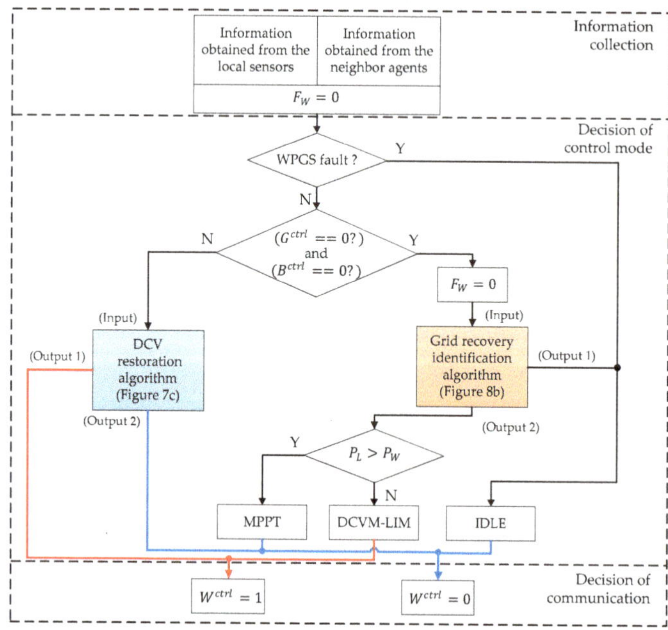

Figure 5. Control strategy of WPGS agent.

If the WPGS has a fault, the WPGS agent switches the operation into IDLE mode, and the variable W^{ctrl} is set to 0 to indicate that the WPGS agent is incapable of controlling the DCV. Otherwise, the WPGS agent determines appropriate control modes according to the supply–demand power relationship to ensure the system power balance. In particular, the MPPT mode is used to extract the maximum power from wind into DC-link. Meanwhile, the DCV control mode by limiting the output power of wind turbine (DCVM-LIM) is implemented to adjust the output power of wind turbine to load demand. Similarly, after the decision of local control mode is given, the information on the ability of the DCV control by the WPGS agent (W^{ctrl}) is simultaneously informed to other agents via communication lines.

Based on the control strategy of the WPGS agent in Figure 5, the local control block to realize all the operating modes is implemented [6].

3.4. Control Strategy of Load Agent

Figure 6 shows the control strategy for local operation of the load agent in MAS-based distributed control approach. In the information collection layer, the information is collected from the local measurement and neighbor agents. Total load demand is obtained from the local measurement and the information on the ability of the DCV control is provided by neighbor agents via the communication lines. If the grid, battery, and WPGS agents can not control the DCV, the operation SHED is activated to remove some less important loads. Similar to the battery and WPGS agents, the load agent is equipped with the DCV restoration algorithm which will be explained in Section 4.1.

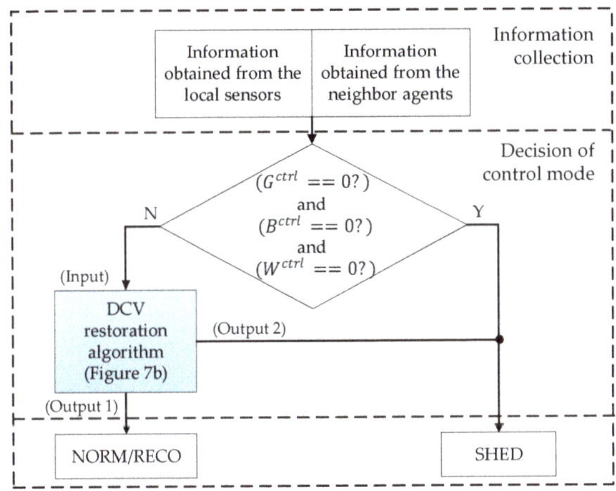

Figure 6. Control strategy of load agent.

4. Proposed Control Strategies under Communication Network Problems

Once the fault is detected in the grid by a detection device, the GO informs the grid fault to the grid agent, and then, the grid agent informs the grid fault to the remaining agents in DCMG. The remaining agents negotiate via communication lines to determine substitute agent which takes over the authority of the DCV regulation.

Unfortunately, in reality, the grid fault cannot be detected instantly by the fault detection device due to the large total response time [29]. Furthermore, delay in transmission lines is unavoidable in the communication-based system. As a result, a significant time delay may exist to recognize the fault. In the worst case of communication failure, the remaining agents may determine operating modes without the information on the grid fault. During such circumstances, the DCV may be increased or decreased rapidly due to the absence of supply–demand power balance since any agents do not serve to regulate the DCV.

As is well known, the DCV stabilization and system power balance should be ensured even in the islanded mode by the DCVM of the remaining agents such as battery or WPGS agent. When the grid is recovered from the grid fault, DCMG operation should be back to grid-connected mode. For this purpose, the GO provides the information on the grid recovery to the grid agent. The grid agent informs the grid recovery to remaining agents by means of the communication network. After receiving the grid recovery information, the battery or WPGS agent can stop the operation DCVM to release the authority of the DCV control to the grid agent. Unfortunately, however, communication problems often prevent the battery or WPGS agent from recognizing the grid recovery instantly. Consequently, these agents keep regulating the DCV with DCVM, which causes a conflict in the DCV control by two voltage control sources at the same time: one by the grid agent and the other by the battery or WPGS agent.

In order to deal with the DCV variation caused by the delay in grid fault detection and communication in the presence of grid fault, the DCV restoration algorithm is adopted in the battery and WPGS agents to restore the DCV rapidly. In addition, to avoid the DCV control by two voltage control sources at the same time under communication problems, a grid recovery identification algorithm is proposed in this paper.

4.1. Control Strategy for Grid Fault Case

Figure 7 shows the DCV restoration algorithms to ensure the DCV stabilization caused by the delay in grid fault detection and communication in the presence of grid fault. Figure 7a shows the detailed DCV restoration algorithm implemented in the battery agent which is mentioned in Figure 4 and Section 3.2. If the DCV is greater than the first threshold level $V_{DC,fault}^{TH1}$, the battery agent operates with IDLE mode. As soon as the DCV is decreased lower than $V_{DC,fault}^{TH1}$, the CPCM starts to restore the DCV by discharging the battery with the maximum discharging power. Simultaneously, the flag F_{B1} is set to 1. As long as the DCV is lower than $V_{DC,fault}^{TH2}$, the battery agent keeps operating with CPCM to restore the DCV as quickly as possible. Once the DCV is increased higher than $V_{DC,fault}^{TH2}$, the battery agent automatically switches the operation into DCVM-DIS to regulate the DCV at the nominal level, setting the flag F_{B2} to 1. With the flags F_{B1} and F_{B2} equal to 1, the battery agent continues to work with DCVM-DIS and the status of the battery agent are transmitted to other agents as shown in Figure 4.

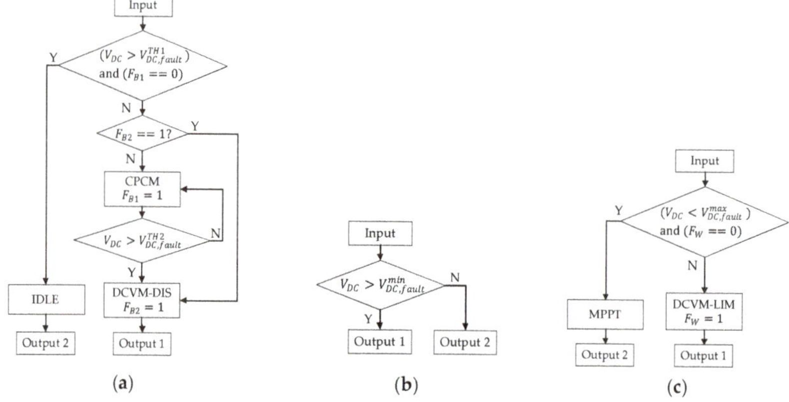

Figure 7. DCV restoration algorithms. (**a**) By battery agent; (**b**) By load agent; (**c**) By WPGS agent.

In some cases, even if the battery supplies the maximum discharging power with CPCM, the DCV can be still decreased due to the deficit power. To avoid the system collapse under such extreme circumstances, the operation SHED is activated in the load agent as shown in Figure 7b. In this algorithm, as soon as the DCV becomes lower than the minimum DCV level $V^{min}_{DC,fault}$, the load agent starts the operation SHED to disconnect less important loads from DC-link.

Figure 7c shows the detailed DCV restoration algorithm in the WPGS agent which is mentioned in Figure 5 and Section 3.3. If the DCV is lower than the maximum level $V^{max}_{DC,fault}$, the WPGS agent works with MPPT mode. When the DCV is increased higher than $V^{max}_{DC,fault}$, the operation DCVM-LIM is triggered to maintain the DCV at the nominal value stably by limiting the output power of the WPGS to total load demand. Simultaneously, the flag F_W is set to 1. Similarly, the status of the WPGS agent is transmitted to other agents as shown in Figure 5.

4.2. Control Strategy for Grid Recovery Case

To improve the DCV stabilization and the DCMG reliability even under communication problems such as delay or failure, a grid recovery identification algorithm is introduced. As can be seen in Figure 3, after receiving the information on the grid recovery from GO, a grid recovery control algorithm in the grid agent is initiated to avoid the DCV control by two voltage control sources at the same time under communication problems.

Figure 8 shows the detailed control strategies for a grid recovery identification algorithm during grid recovery. First, once the grid is recovered from the fault, the grid agent operates with special current control mode (SCCM) as shown in Figure 8a, in which the special current pattern is generated in a square wave with a frequency of f_G on the DC-link power line. Since periodic signals are superimposed on DC values, these signals can be easily detected with simple signal processing schemes irrespective of chosen frequency and amplitude. In this paper, the special current pattern can be detected in the battery or WPGS agent by monitoring the battery current I_B or the q-axis current of WPGS $I_{W,q}$. Figure 8b shows the grid recovery identification algorithm to detect the special current pattern generated by the grid agent, which consists of a high pass filter, zero-crossing detection, and frequency calculation. By comparing the detected frequency f'_G with the generated frequency f_G by the grid agent, the grid recovery can be effectively identified by the battery and WPGS agents. After the grid recovery is recognized, the battery or WPGS agent stops the operation DCVM, which causes the DCV variation. By monitoring this DCV variation with $V^{min}_{DC,\ reco}$ and $V^{max}_{DC,\ reco}$, the grid agent switches the operation SCCM into DCVM-REC or DCVM-INV to regulate the DCV as shown in Figure 8a. The flags F_{G2} and F_{G3} are used to denote the corresponding operating mode. The output information of the grid recovery identification algorithm is fed back to Figures 4 and 5 to determine the operating modes of the battery and WPGS.

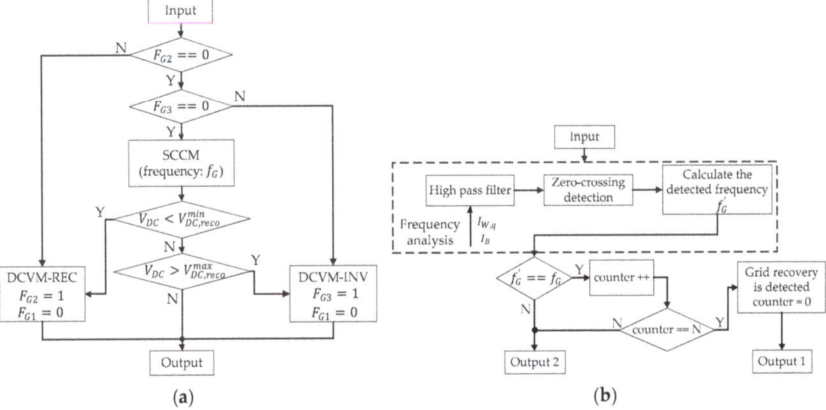

Figure 8. Control strategies for reliability improvement of DCMG during grid recovery. (**a**) Special current pattern generation by grid agent; (**b**) Grid recovery identification algorithm of battery and WPGS agents by frequency detection.

Figure 9 shows the illustration to validate the proposed grid recovery identification algorithm by frequency detection in the battery agent. As can be seen in Figure 9c, with the special current pattern generated in square wave with a frequency of f_G by the grid agent, the battery current I_B includes two components: low frequency component for the normal operation and high frequency component due to the special current pattern by the grid agent. By using high pass filter and zero-crossing detection, the generated frequency f_G can be effectively detected to recognize the grid recovery.

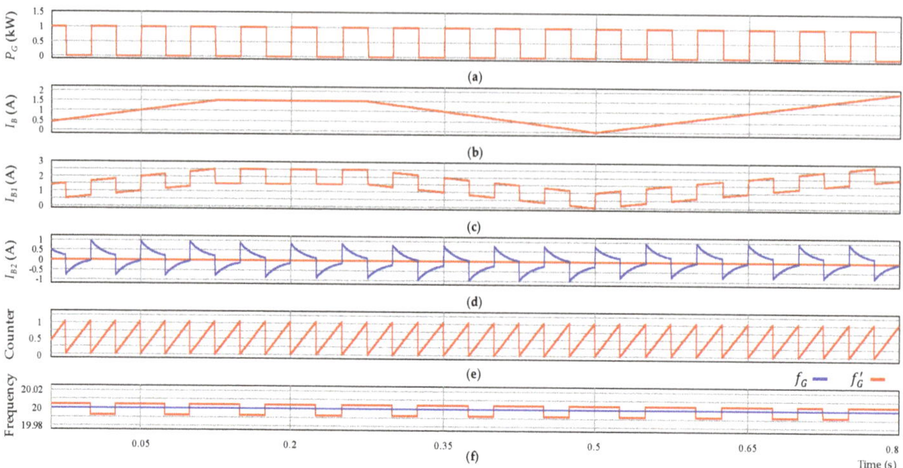

Figure 9. Illustration of grid recovery identification algorithm by frequency detection. (**a**) Output power of grid agent; (**b**) Battery current without special current pattern; (**c**) Battery current with special current pattern; (**d**) Output of high pass filter; (**e**) Output of zero-crossing detection; (**f**) Generated frequency by the grid agent and detected frequency by grid recovery identification algorithm.

5. Simulation Results

In order to verify the feasibility of the PMS and the reliability of the proposed control strategies under communication problems, the simulations based on the PSIM are conducted. Table 2 lists the system parameters used for simulations. The detailed mathematical models of three converters used for the grid agent, WPGS agent, and battery agent can be found in [30–32]. Simulation results are presented in three cases which are the grid-connected case, the islanded case, and the case of communication problems.

Table 2. System parameters of DCMG.

Unit	Parameter	Symbol	Value
Grid operator	Maximum power supplied from grid to DCMG	$P_{G,rec}^{max}$	2 kW
	Maximum power injected to grid from DCMG	$P_{G,inv}^{max}$	2 kW
Grid agent	Grid voltage	V_G^{rms}	220 V
	Grid frequency	F_G	60 Hz
	Frequency for SCCM	f_G	20 Hz
	Transformer Y/Δ	T	380/220 V
	Local controller: Digital signal processor (DSP) TMS320F28335	-	-

Table 2. *Cont.*

Unit	Parameter		Symbol	Value
WPGS agent	Induction machine	Rated power	P_{Ind}^{rated}	3 kW
		Rated voltage	V_{Ind}^{rated}	380 V
		Rated speed	n_{Ind}^{rated}	1425 min^{-1}
		Power factor	$\cos\varphi$	0.85
	Permanent magnet synchronous generator (PMSG)	Stator resistance	R_S	0.64 Ω
		dq-axis inductance	L_{dq}	0.82 mH
		Number of pole pairs	P	6
		Inertia	J	0.111 kgm^2
		Flux linkage	ψ	0.18 Wb
		Rated power	P_{Gen}^{rated}	3 kW
		Maximum speed	n_{Gen}^{max}	3200 rpm
		Voltage constant	K_e	147 V/Krpm
		Predefined level for frequency detection in grid recovery identification algorithm	N	11
		Local controller: DSP TMS320F28335	-	-
Battery agent	Maximum discharging power		$P_{B,dis}^{max}$	2 kW
	Maximum voltage		V_B^{max}	265 V
	Maximum SOC		SOC^{max}	90 %
	Minimum SOC		SOC^{min}	10 %
	Rated capacity		C	30 Ah
	Predefined level for frequency detection in grid recovery identification algorithm		N	11
	Local controller: DSP TMS320F28335		-	-
Load agent	Power of load 1		P_{L1}	0.5 kW
	Power of load 2		P_{L2}	1 kW
	Power of load 3		P_{L3}	2 kW
	Priority level: load 1 > load 2 > load 3		-	-
	Local controller: DSP TMS320F28335		-	-
DC-link	Nominal voltage		V_{DC}^{nom}	400 V
	Maximum DCV used for grid fault		$V_{DC,fault}^{max}$	420 V
	First threshold level for DCV used for grid fault		$V_{DC,fault}^{TH1}$	370 V
	Second threshold level for DCV used for grid fault		$V_{DC,fault}^{TH2}$	390 V
	Minimum DCV used for grid fault		$V_{DC,fault}^{min}$	360 V
	Maximum DCV used for grid recover		$V_{DC,reco}^{max}$	410 V
	Minimum DCV used for grid recover		$V_{DC,reco}^{min}$	390 V
	Capacitance		C_{DC}	4 mF

5.1. Grid-connected Case

In the grid-connected case, the grid agent primarily controls the DCV to maintain the system power balance under variations of wind power and load demand. Table 3 lists the operation conditions used for the simulation tests.

Table 3. Operation conditions for simulation test in grid-connected mode.

	Operation Conditions	Time (s)
1	Load 3 is switched in.	1
2	Wind power gradually decreases from 0.5 kW to 0.	1.5
3	Wind power increases from 0 to 2 kW.	2
4	Wind power increases from 2 kW to 4 kW.	2.5
5	Load 2 and load 3 are switched out.	3
6	Battery fault occurs.	3.5

Figure 10 shows the simulation results for the grid-connected mode with the operation conditions given in Table 3. It is assumed that the load 1 and load 2 are initially connected to the DC-link, consuming the power of 1.5 kW. Although the WPGS works with the MPPT mode, the output power of the WPGS is not sufficient to supply the load. To deal with this power imbalance, the grid agent operates with DCVM-REC to compensate the deficit power by supplying more power from the grid to the DC-link. In this instant, the DCV is controlled to the nominal value of 400 V by the grid agent and the battery agent is in IDLE mode.

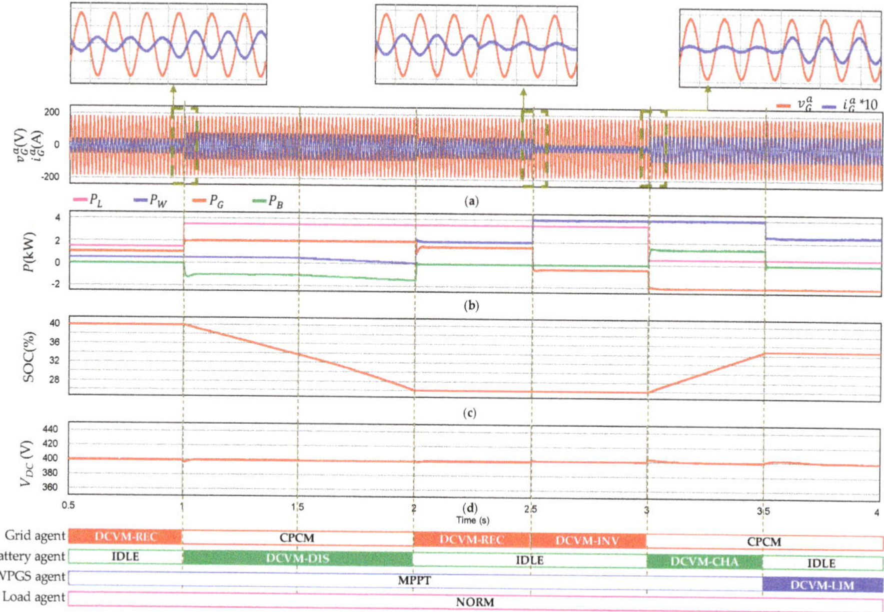

Figure 10. Simulation results for grid-connected mode. (**a**) a-phase grid voltage and current; (**b**) Output power of each agent; (**c**) Battery *SOC*; (**d**) DCV.

At t = 1 s, the load 3 is switched in, which results in an increase of total load demand. To ensure the system power balance, the grid agent further increases the supplied power from the grid to the DC-link. However, since the supplied power is limited by the maximum power of 2 kW, the grid agent switches the operation into CPCM, and sends the data ($G^{ctrl} = 0$) to other agents. As soon as receiving the data, the battery agent switches its operation from IDLE to DCVM-DIS to maintain the system power balance by regulating the DCV.

From t = 1.5 s, as the wind power gradually decreases from 0.5 kW to zero, the battery agent gradually increases the discharging power to compensate for the change of wind power as can be seen in Figure 10b.

At t = 2 s, the wind power suddenly changes from zero to 2 kW. Although the wind power is still lower than the load demand of 3.5 kW, the grid agent can control the DCV with DCVM-REC since the supply power of 1.5 kW from the grid to DC-link is within the maximum power level. As a result, the battery agent moves to IDLE mode after receiving the data ($G^{ctrl} = 1$) from the grid agent.

At t = 2.5 s, the wind power is further increased to 4 kW, which is higher than load demand. Then, the grid agent switches the operation DCVM-REC into DCVM-INV to transfer the surplus power from the DCMG to the grid.

When the load 2 and load 3 are switched out at t = 3 s, the grid agent tries to inject more power to the grid to maintain the supply–demand power balance. However, since the exchange power exceeds the maximum level assigned by GO, the grid agent can transfer only a portion of surplus power to the grid by CPCM, transmitting the data ($G^{ctrl} = 0$) to other agents in DCMG. After receiving the data via communication, the battery agent switches the operation into DCVM-CHA to absorb the remaining surplus power.

At t = 3.5 s, when the battery agent is incapable of regulating the DCV due to battery fault, the battery agent changes the operation into IDLE mode, and transmits the data ($B^{ctrl} = 0$) to other agents. Because both the grid and battery agents cannot control the DCV in this instant, the WPGS agent intervenes with DCVM-LIM to maintain the output power of WPGS to load power. It is shown in Figure 10d that the DCV can be always regulated stably at the nominal value in all the operating conditions regardless of the conditions of the battery, wind power, and load demand.

5.2. Islanded Case

Since the grid agent is incapable of controlling the DCV in the islanded case, the DCV control is achieved by the negotiation of remaining agents in DCMG such as the battery, WPGS, and load agents. The operation conditions used for the simulation tests are listed in Table 4.

Table 4. Operation conditions for simulation test in islanded mode.

	Operation Conditions	Time (s)
1	Grid fault occurs.	1
2	Wind power decreases from 2 kW to 1 kW.	1.5
3	Battery *SOC* reaches the minimum level.	2
4	Wind power increases from 1 kW to 2 kW.	2.5
5	Load 2 is reconnected.	3
6	Grid is recovered.	3.5

The simulation results for the islanded mode according to the operation conditions in Table 4 are shown in Figure 11. Initially, the system power balance is guaranteed by the grid agent with DCVM-REC. Meanwhile, the WPGS works in the MPPT mode to produce 2 kW, the battery is in IDLE mode and all loads are fed.

When the grid fault occurs at t = 1 s, the system operation is changed to the islanded mode. After receiving the fault information from GO, the grid agent switches the operation to IDLE mode, and simultaneously sends the data ($G^{ctrl} = 0$) to other agents. Because the wind power of 2 kW is smaller than the load demand of 3.5 kW, the battery agent rapidly enters DCVM-DIS to compensate for the deficit power.

At t = 1.5 s, as the wind power suddenly drops to 1 kW, the battery agent tries to increase the discharging power to maintain the supply–demand power balance. However, due to the discharging power limit of 2 kW, the battery agent starts CPCM. In this instant, the battery agent cannot maintain the system power balance as well as control the DCV. After recognizing that both the grid and battery agents are not able to control the DCV by communication information, the load agent activates SHED to disconnect the load 3 which has the lowest priority. A time delay T_{shed} is used in SHED to avoid undesirable load disconnection caused by noise. After SHED, the battery can return to DCVM-DIS to continue the DCV regulation.

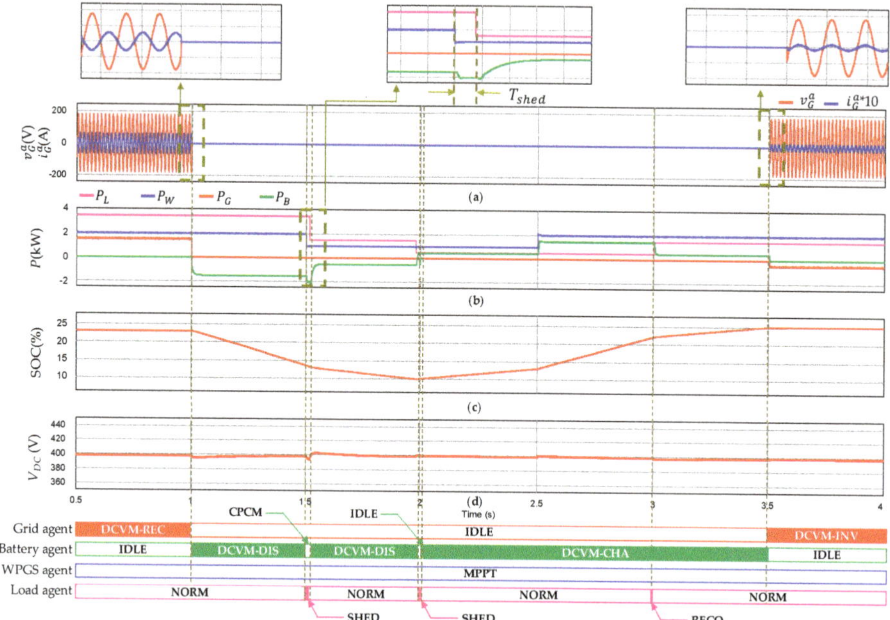

Figure 11. Simulation results for islanded mode. (**a**) *a*-phase grid voltage and current; (**b**) Output power of each agent; (**c**) Battery *SOC*; (**d**) DCV.

At t = 2 s, as the battery *SOC* reaches the minimum level of 10%, the battery agent switches the operation into IDLE mode, and the load agent executes the operation SHED again to disconnect the load 2. With the disconnection of the load 2, the wind power is sufficient to supply the remaining load power of 0.5 kW. Then, the battery agent starts DCVM-CHA to absorb the surplus power in DCMG.

At t = 2.5 s, as the wind power increases from 1 kW to 2 kW, the DCV is continuously regulated with DCVM-CHA by the battery agent. This operation is still maintained when the load 2 is reconnected to the DC-link at t = 3 s.

The grid is recovered at t = 3.5 s. Assuming that there are no communication problems, the grid agent informs the grid recovery to other agents by sending the data ($G^{ctrl} = 1$). Then, the battery agent switches the operation into IDLE mode while the DCV is regulated by the grid agent with DCVM-INV. From Figure 11, it can be concluded that the DCV is regulated stably at the nominal value in all the operating conditions by using the MAS-based distributed control strategy under the variations of wind power and load demand in the islanded case.

5.3. Case of Communication Problems

In this Section, the simulation results are presented to validate the effectiveness of the proposed control schemes under communication problems. The results of the DCV restoration scheme to deal with the delay in grid fault detection and communication are shown in Figure 12, Figure 13, and Figure 14. The results for the grid recovery detection under communication failure are presented in Figures 15 and 16.

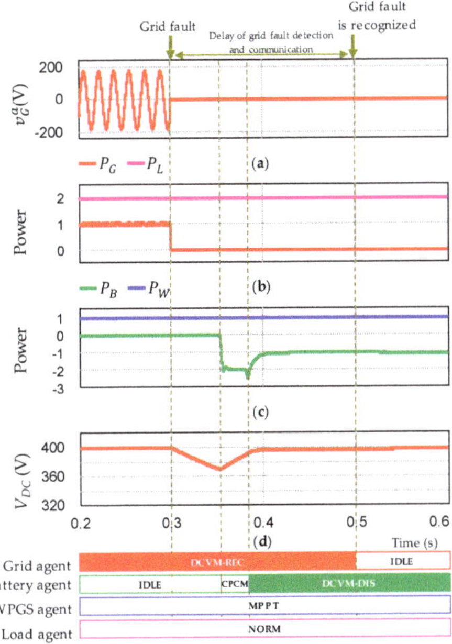

Figure 12. Simulation results of the proposed DCV restoration by battery agent without SHED. (**a**) *a*-phase grid voltage; (**b**) Output power of grid agent and total load power; (**c**) Output power of battery agent and WPGS agent; (**d**) DCV.

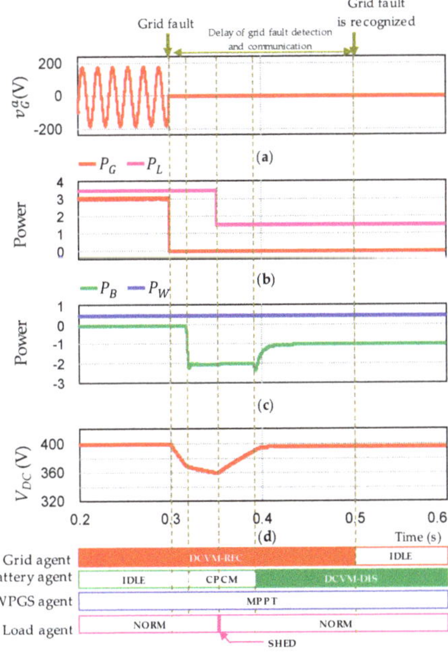

Figure 13. Simulation results of the proposed DCV restoration by battery agent with SHED. (**a**) *a*-phase grid voltage; (**b**) Output power of grid agent and total load power; (**c**) Output power of battery agent and WPGS agent; (**d**) DCV.

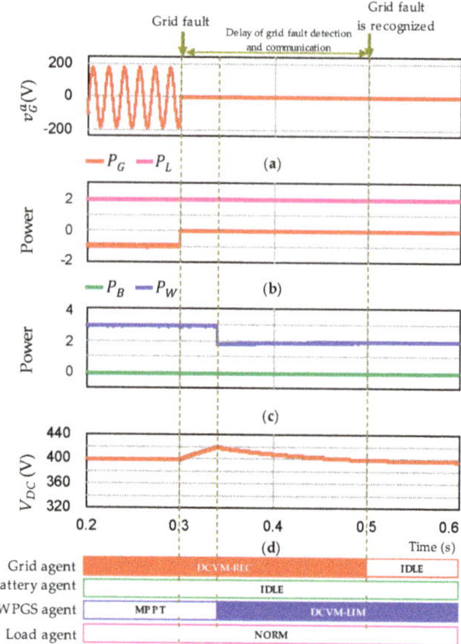

Figure 14. Simulation results of the proposed DCV restoration by WPGS agent. (**a**) *a*-phase grid voltage; (**b**) Output power of grid agent and total load power; (**c**) Output power of battery agent and WPGS agent; (**d**) DCV.

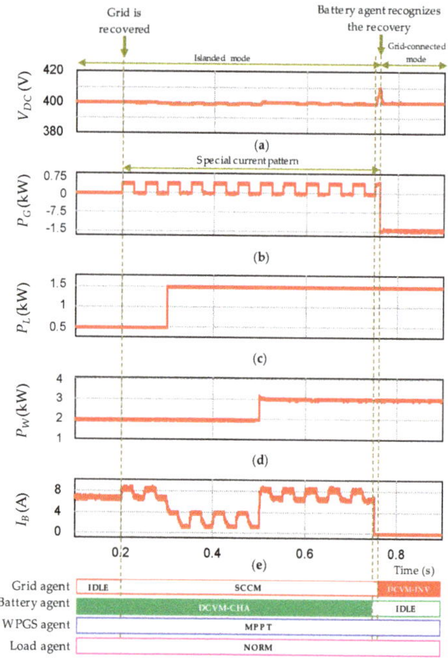

Figure 15. Simulation results for the grid recovery identification under communication failure by battery agent. (**a**) DCV; (**b**) Output power of grid agent; (**c**) Load power; (**d**) Output power of WPGS; (**e**) Battery current.

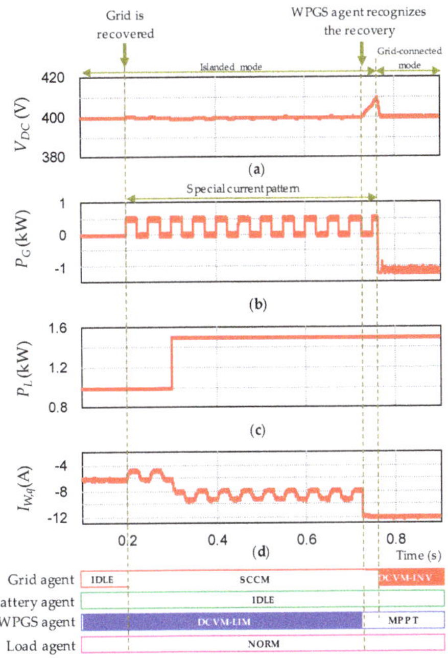

Figure 16. Simulation results for the grid recovery identification under communication failure by WPGS agent. (**a**) DCV; (**b**) Output power of grid agent; (**c**) Load power; (**d**) WPGS current.

Figure 12 shows the simulation results of the proposed DCV restoration algorithm by the battery agent without SHED in the presence of delay in grid fault detection and communication. Before the grid fault, the DCV control and system power balance is taken over by the grid agent. Meanwhile, the battery agent is in the IDLE mode and the WPGS agent works with the MPPT mode. It is assumed that the grid has a fault at t = 0.3 s and GO and all the system agents cannot instantly recognize it due to delay in grid fault detection and communication. In this condition, the grid agent still works with DCVM-REC while the battery is in IDLE mode. As a result, the DCV decreases rapidly because of power imbalance as can be seen in Figure 12c. However, according to the proposed DCV restoration algorithm in Figure 7, the battery agent executes CPCM and DCVM-DIS in turn to restore the DCV to the nominal value of 400 V. When the grid fault is detected at t = 0.5 s, the operation of the grid agent is switched into IDLE, and the battery agent continuously controls the DCV by DCVM-DIS.

Figure 13 shows the simulation results of the proposed DCV restoration algorithm by the battery agent with SHED under the delay in grid fault detection and communication. The algorithm operates similarly to Figure 12. Only the difference is to apply SHED when the DCV cannot be restored with CPCM. It is shown in Figure 13c that the DCV still decreases even when the battery agent operates with CPCM. As soon as the DCV reaches V_{fault}^{min}, the load agent triggers SHED to support the DCV restoration. Once the DCV ceases decreasing, the battery agent executes CPCM and DCVM-DIS in turn to restore the DCV completely to the nominal value of 400 V.

Figure 14 shows the simulation results of the proposed DCV restoration algorithm by the WPGS agent in the presence of delay in grid fault detection and communication. Before the grid fault, the WPGS agent operates with the MPPT mode, the battery is in IDLE, and the DCV is regulated to 400 V with DCVM-INV by the grid agent. As before, it is assumed that the grid has a fault at t = 0.3 s, and GO and all the system agents cannot instantly recognize it due to delay in grid fault detection and communication. This results in the rapid increase of the DCV due to supply–demand power imbalance. However, even in this case, the WPGS agent can effectively restore the DCV with DCVM-LIM by using the proposed DCV restoration algorithm in Figure 7. Once the grid fault is detected at t = 0.5 s, the operation of grid agent is switched into IDLE, and the WPGS agent continuously controls the DCV by DCVM-LIM.

Figures 15 and 16 shows the simulation results for the grid recovery identification under communication failure by the battery agent and WPGS agent, respectively. In Figure 15, the DCV is regulated with DCVM-CHA by the battery agent before the grid is recovered. Even if the grid is recovered at t = 0.2 s, the battery agent cannot recognize it under the communication failure, keep operating with DCVM-CHA. To prevent this situation, the grid agent activates the SCCM to inject a special current pattern (square wave with frequency of 20 Hz) to DC-link. Based on the proposed grid recovery identification algorithm in Figure 8b, the battery current waveform is analyzed to identify the grid recovery by detecting current pattern. The grid recovery can be detected in 0.6 s to ensure the reliable detection under the effect of wind power and load demand variations as can be shown in Figures 15c and 15d. When the battery agent recognizes the grid recovery, the battery agent stops the operation DCVM-CHA to release the authority of DCV control to the grid agent. As a result of ceasing the DCV control, the DCV varies instantly according to supply–demand power balance in DCMG. By sensing this DCV variation, the grid agent can switch the operation SCCM into DCVM-INV to restart the DCV regulation under the communication failure.

In Figure 16, the DCV is regulated with DCVM-LIM by the WPGS agent before grid is recovered. By using the similar procedure, the WPGS can identify the grid recovery by the proposed grid recovery identification algorithm. Then, the WPGS agent switches the operation DCVM-LIM into MPPT to release the authority of DCV control to the grid agent. As a result, it is confirmed from the simulation results that the grid recovery can be detected by using the proposed scheme even under the communication failure irrespective of the variations of wind and load power. In addition, the DCV can be stably regulated without any conflict in the DCV control by two voltage control sources at the same time.

6. Experimental Results

In order to verify the feasibility of the PMS and the reliability of the proposed control strategies in practice, the experiments have been conducted using a laboratory testbed system. Figure 17 shows the configuration of a laboratory DCMG system which consists of three power converters for the connections with the grid, battery, and WPGS. For the grid agent, a bidirectional AC/DC converter which is connected to 3-phase grid through Y-Δ transformer and LCL filters, is employed to exchange the power between the grid and DC-link. In the WPGS agent, a PMSG coupled with an induction machine is used to emulate the wind turbine. The developed power of the WPGS agent is adjusted by controlling the rotating speed of the induction machine through the induction machine drive system. The output power of PMSG is transferred to DC-link by a unidirectional AC/DC converter. In this study, the DSP TMS320F28335 is used to control the converters. The detailed information of the experimental rig can be seen in Figure 2.

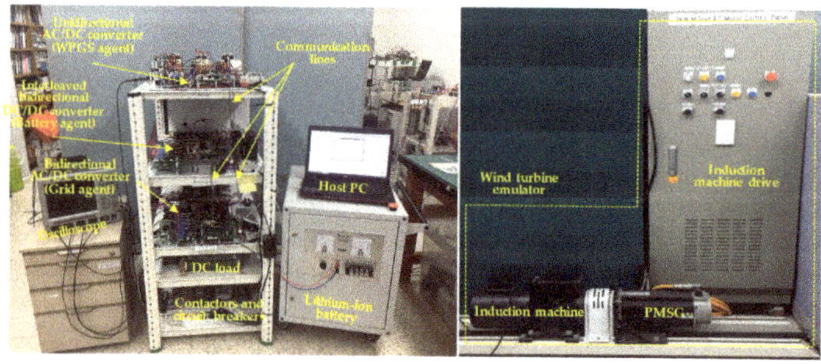

Figure 17. Configuration of the experimental DCMG system.

6.1. Grid-connected Case

Figure 18 shows the experimental results for system operation in the grid-connected mode by using MAS-based distributed control under the variations of load and wind power. Figure 18a shows DCMG operation under load variation. Initially, the grid agent operates with DCVM-REC to supply the power of 1.2 kW from the grid to DCMG. As the load demand is increased suddenly from 1.2 kW to 2.8 kW, the grid agent increases supplied power from the grid to DCMG to ensure supply–demand power balance. Similar to the simulation result in Figure 10, as the grid supplied power reaches the maximum exchange power of 2 kW, the grid agent switches the operation into CPCM, and sends the data ($G^{ctrl} = 0$) to other agents. After receiving the data, the battery agent switches the operation into DCVM-DIS to compensate for the deficit power.

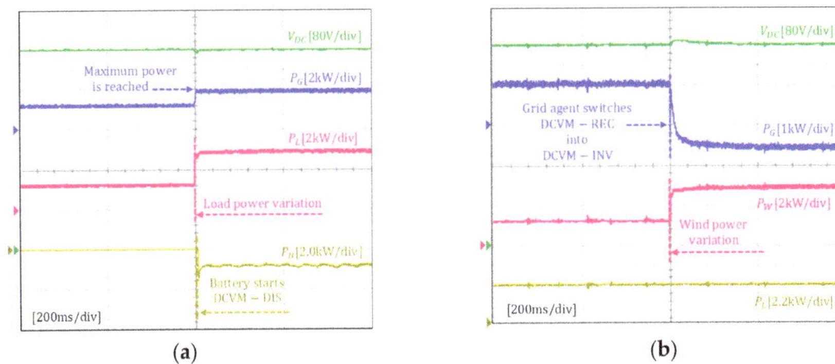

Figure 18. Experimental results for grid-connected mode. **(a)** Load variation; **(b)** Mode transition of grid agent from DCVM-REC to DCVM-INV.

Figure 18b shows the system behavior under wind power variation. Before the variation, the WPGS agent works with the MPPT mode, and the grid agent operates with DCVM-REC to supply the power from the grid to DCMG. When the wind power increases beyond the load demand, the grid agent switches the operation DCVM-REC into DCVM-INV to inject the surplus power to the grid. As can be seen from Figure 18, the DCV is stably regulated at 400 V by using the MAS-based distributed control in spite of the variations of load demand and wind power. In these results, Figure 18a corresponds to the instant of t = 1 s in the simulation result of Figure 10, and Figure 18b corresponds to the instant of t = 2.5 s in Figure 10.

6.2. Islanded Case

Figure 19 shows the experimental results for MAS-based distributed control during the transition from the grid-connected to the islanded mode, and vice versa. Figure 19a illustrates DCMG operation for the transition from the grid-connected to islanded mode due to grid fault. Before the grid fault occurs, the system power balance is well maintained by the grid agent. At the instant of grid fault, the battery agent starts the operation DCVM-DIS to compensate for the deficit power caused by the absence of grid. Consequently, the DCV is reliably maintained at 400 V. Figure 19b shows the results for the transition from the islanded to grid-connected mode. Before the grid is recovered, the battery agent maintains the DCV at the nominal value by DCVM-CHA to absorb the surplus power of 0.8 kW. Without communication problems, as soon as the grid is recovered, the battery agent stops the operation DCVM-CHA. Instead, the grid agent activates DCVM-INV to regulate the DCV stably by injecting the surplus power to the grid. These experimental results correspond to the simulation results at t = 1 s and t = 3.5 s in Figure 11 and validate the feasibility of the PMS by using MAS-based distributed control.

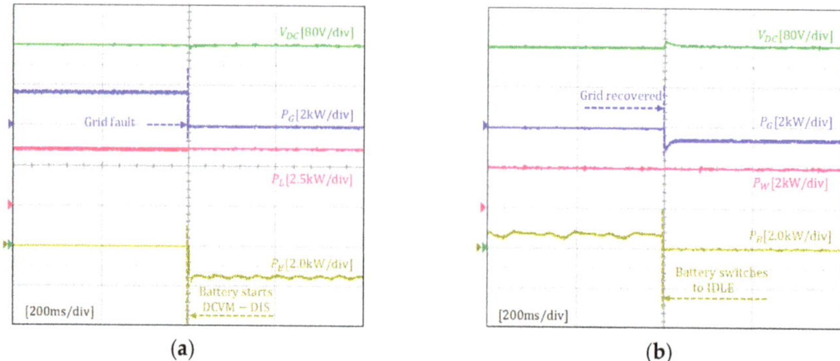

(a) (b)

Figure 19. Experimental results for islanded mode. (**a**) Transition from grid-connected to islanded mode; (**b**) Transition from islanded mode to grid-connected mode.

6.3. Case of Communication Problems

This Section presents the experimental results of the proposed DCV restoration and grid recovery identification.

Figures 20a and 20b show the results of the proposed DCV restoration under the delay in grid fault detection and communication by battery agent without SHED and with SHED, respectively. In Figure 20a, the DCV is controlled by the grid agent while the battery agent is in IDLE mode before the grid fault. At the instant of the grid fault, according to the proposed DCV restoration algorithm in Figure 7, the battery agent executes CPCM and DCVM-DIS in turn to restore the DCV, which coincides well with Figure 12 in the simulation. Similar to Figure 13 in the simulation, the proposed DCV restoration algorithm executes CPCM by the battery agent, SHED by the load agent, and DCVM-DIS by the battery agent in turn to regulate the DCV stably. Figure 20c shows the results of the proposed DCV restoration by the WPGS agent under the delay in grid fault detection and communication. Before the grid fault, the DCV is regulated at 400 V with DCVM-REC by the grid agent. When the grid has a fault and the battery cannot be used to control the DCV, the WPGS agent intervenes to take a role of DCV regulation as shown in Figure 20c.

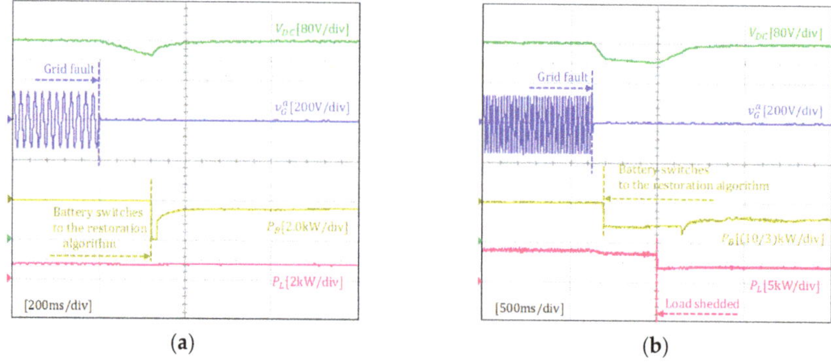

(a) (b)

Figure 20. *Cont.*

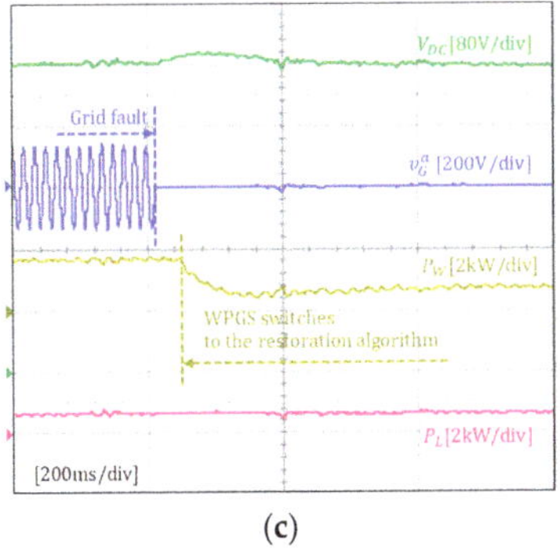

(c)

Figure 20. Experimental results of the proposed DCV restoration. (**a**) By battery agent without SHED; (**b**) By battery agent with SHED (**c**) by WPGS agent.

All the experimental results are matched well with the simulations in Figure 12 through Figure 14, which confirms the feasibility of the DCV restoration algorithm.

Figure 21 shows the experimental results of the grid recovery identification scheme under the communication failure by the battery agent in Figure 21a and by the WPGS agent in Figure 21b, respectively. In Figure 21a, the DCV is regulated with DCVM-CHA by the battery agent before the grid recovery. Once the grid is recovered, the battery agent can detect the grid recovery by using the proposed grid recovery identification scheme and switches the operation into IDLE mode. Instead, the DCV regulation is taken over by the grid agent with DCVM-INV. In Figure 21b, if the grid is recovered when the WPGS agent controls the DCV, the WPGS agent identifies the grid recovery, and switches the operation into the MPPT mode. Similarly, the DCV regulation is taken over by the grid agent from this instant. These results also accord well with the simulation results in Figures 15 and 16.

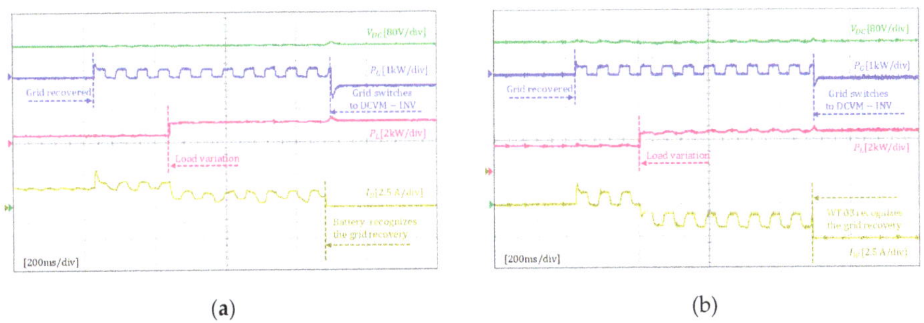

(**a**) (**b**)

Figure 21. Experimental results for the grid recovery identification under communication failure. (**a**) By battery agent; (**b**) By WPGS agent.

It is confirmed that the grid recovery can be effectively detected by using the proposed scheme even under communication failure in the presence of the disturbance such as load power variation. As a result, the DCV can be well regulated to the nominal value without any conflict in the DCV control by two voltage control sources.

In case of the grid fault, the DCV can be restored to the nominal value in 0.1~0.2 s in the simulation results of Figure 12–14, and in 0.2~0.5 s in the experimental results of Figure 20, depending on the communication speed, the converter control dynamics, and DC link capacitance. In both the simulation and experiment, the maximum variation of the DCV during restoration process is obtained as $\Delta V_{DC}/V_{DC} = \frac{40}{400} \cdot 100\% = 10\%$ as shown in Figures 13 and 20b.

In case of the grid recovery, the grid recovery can be detected within 0.6 s even under the communication failure as shown in both the simulation and experiment of Figure 15, Figure 16, and Figure 21. The maximum variation of the DCV during detection process is $\Delta V_{DC}/V_{DC} = \frac{10}{400} \cdot 100\% = 2.5\%$.

7. Conclusions

For the purpose of enhancing the reliability of DCMG and ensuring the system power balance under various conditions, this paper has presented an improved PMS for MAS-based distributed control of DCMG under communication network problems. The main contributions of this paper can be summarized as follows:

i) A MAS-based distributed control technique has been developed for PMS of DCMG including four agents of the grid, WPGS, battery, and load. In the PMS, by investigating the information obtained from both the local measurements and the neighboring agents via communication lines, all the agents determine not only optimal operating modes for local controllers but also the proper information to be transmitted in order to guarantee the system power balance under various conditions. By using this control scheme, the drawbacks of the existing centralized and decentralized controls are eliminated. In particular, the control mode of the agents can be determined locally without CC, which contributes to avoid the common single point of failure as in the centralized control. Since each agent is controlled by its own local controller, the computational burden is shared by local controllers. In addition, the problem related to communication failure can be avoided by using the proposed reliability improvement algorithm. As compared to the decentralized control, the problems related to the lack of exchanged information can be solved by means of communication network.

ii) In order to deal with the system power imbalance caused by the delay in grid fault detection and communication, the DCV restoration algorithm is proposed in this paper. By using this scheme, as soon as abnormal level of the DCV is detected, the DCV can be restored stably to the nominal value irrespective of unbound time delay even under the communication network problems, which contributes to prevent the system malfunction and collapse of DCMG.

iii) To recognize the grid recovery reliably even under communication failure, the grid recovery identification algorithm is introduced in this study. By detecting the frequency of the current pattern on the DC-link, the battery and WPGS agents can effectively identify the grid recovery even in the presence of communication failure. Therefore, the conflict in the DCV control by different voltage control sources can be avoided.

In order to validate the feasibility of the PMS and the proposed schemes, the simulations based on PSIM and the experiments based on the laboratory DCMG testbed have been carried out. In the experimental setup, the WPGS is emulated by coupled PMSM-induction machine. Comprehensive simulation and experimental results have proven the usefulness of the PMS and the proposed reliability improvement schemes under various conditions, especially, under communication network problem. As a result, the proposed DCV restoration and grid recovery identification schemes can contribute to enhance the reliability of DCMG system.

Author Contributions: T.V.N. and K.-H.K. conceived the main concept of the control structure and developed the entire system. T.V.N. carried out the research and analyzed the numerical data with guidance from K.-H.K. T.V.N. and K.-H.K. collaborated in the preparation of the manuscript. All authors have read and agreed to the published version of the manuscript.

Funding: This study was supported by the Advanced Research Project funded by the SeoulTech (Seoul National University of Science and Technology).

Conflicts of Interest: The authors declare no conflict of interest.

Abbreviation

AC	Alternating current
ACMG	AC microgrid
CC	Central controller
CPCM	Constant power control mode
CVCM	Constant voltage control mode
DC	Direct current
DCMG	DC microgrid
DCV	DC-link voltage
DCVM	DC-link voltage control mode
DCVM-CHA	DC-link voltage control mode by battery charging
DCVM-DIS	DC-link voltage control mode by battery discharging
DCVM-INV	DC-link voltage control mode by inverter mode of grid agent
DCVM-LIM	DC-link voltage control mode by limiting output power of WPGS
DCVM-REC	DC-link voltage control mode by converter mode of grid agent
DG	Distributed generation
ESS	Energy storage system
GO	Grid operator
MAS	Multiagent system
MPPT	Maximum power point tracking
PMS	Power management strategy
PMSG	Permanent magnet synchronous generator
RECO	Load reconnection
RES	Renewable energy source
SCCM	Special current control mode
SHED	Load shedding
SOC	State of charge
WPGS	Wind power generation system
NORM	Normal load state
C	Battery rated capacity
$\cos\varphi$	Power factor of induction machine
C_{DC}	DC-link capacitance
F_{B1}, F_{B2}	Flags in control strategy of battery agent
f_G	Frequency of special current pattern for grid identification algorithm
f'_G	Detected frequency for grid identification algorithm
F_{G1}, F_{G2}, F_{G3}	Flags in control strategy of grid agent
F_W	Flag in control strategy of WPGS agent
i^a_G	a-phase grid current
I_B	Battery current
I_{B1}	Battery current with special current pattern
I_{B2}	Output of high pass filter
$G^{ctrl}, B^{ctrl}, W^{ctrl}$	Indications of DCV control ability by grid, battery, and WPGS agents
G^{state}	State of grid
K_e	Voltage constant of PMSG
n^{max}_{Gen}	Rated speed of wind turbine generator
n^{rated}_{Ind}	Rated speed of induction machine
P_B	Output power of battery agent
$P^{max}_{B,\,cha}, P^{req}_{B,\,cha}$	Maximum and required battery charging power
$P^{max}_{B,\,dis}, P^{req}_{B,\,dis}$	Maximum and required battery discharging power
P_G	Output power of grid agent
$P^{max}_{G,inv}, P^{req}_{G,inv}$	Maximum and required power injected to the grid from DCMG
$P^{max}_{G,rec}, P^{req}_{G,rec}$	Maximum and required power supplied from the grid to DCMG
P^{rated}_{Gen}	Rated power of PMSG
P^{rated}_{Ind}	Rated power of induction machine
$P_L, P_{L1}, P_{L2}, P_{L3}$	Total load power, and power of load 1, load 2, load 3

P_W	Output power of WPGS agent
P_{W-L}	Supply-demand power relationship
SOC^{max}, SOC^{min}	Maximum and minimum battery state of charge
T_{shed}	Time delay for load shedding
v_G^a	a-phase grid voltage
V_B, V_B^{max}	Battery voltage and maximum battery voltage
V_{DC}, V_{DC}^{nom}	DCV, nominal DCV
$V_{DC,fault}^{max}$, $V_{DC,fault}^{TH1}$, $V_{DC,fault}^{TH2}$, $V_{DC,fault}^{min}$	DCV thresholds for grid fault case
$V_{DC,reco}^{max}$, $V_{DC,reco}^{min}$	DCV thresholds for grid recovery case
V_{Ind}^{rated}	Rated voltage of induction machine

References

1. Hatziargyriou, N.; Asano, H.; Iravani, R.; Marnay, C. Microgrids. *IEEE Power Energy Mag.* **2007**, *5*, 78–94. [CrossRef]
2. Farrokhabadi, M.; Koenig, S.; Canizares, C.A.; Bhattacharya, K.; Leibfried, T. Battery Energy Storage System Models for Microgrid Stability Analysis and Dynamic Simulation. *IEEE Trans. Power Syst.* **2018**, *33*, 2301–2312. [CrossRef]
3. Mohammadi, F.; Nazri, G.-A.; Saif, M. A Bidirectional Power Charging Control Strategy for Plug-in Hybrid Electric Vehicles. *Sustainability* **2019**, *11*, 4317. [CrossRef]
4. Sanjeev, P.; Padhy, N.P.; Agarwal, P. Peak Energy Management Using Renewable Integrated DC Microgrid. *IEEE Trans. Smart Grid* **2018**, *9*, 4906–4917. [CrossRef]
5. Mohammadi, F.; Nazri, G.-A.; Saif, M. A New Topology of a Fast Proactive Hybrid DC Circuit Breaker for MT-HVDC Grids. *Sustainability* **2019**, *11*, 4493. [CrossRef]
6. Nguyen, T.V.; Kim, K.-H. Power Flow Control Strategy and Reliable DC-Link Voltage Restoration for DC Microgrid under Grid Fault Conditions. *Sustainability* **2019**, *11*, 3781. [CrossRef]
7. El-Hendawi, M.; Gabbar, H.A.; El-Saady, G.; Ibrahim, E.-N.A. Control and EMS of a Grid-Connected Microgrid with Economical Analysis. *Energies* **2018**, *11*, 129. [CrossRef]
8. Xu, L.; Chen, D. Control and Operation of a DC Microgrid with Variable Generation and Energy Storage. *IEEE Trans. Power Deliv.* **2011**, *26*, 2513–2522. [CrossRef]
9. Dragičević, T.; Lu, X.; Vasquez, J.C.; Guerrero, J.M. DC Microgrids-Part I: A Review of Control Strategies and Stabilization Techniques. *IEEE Trans. Power Electron.* **2016**, *31*, 4876–4891.
10. Dehkordi, N.M.; Sadati, N.; Hamzeh, M. Fully Distributed Cooperative Secondary Frequency and Voltage Control of Islanded Microgrids. *IEEE Trans. Energy Convers.* **2017**, *32*, 675–685. [CrossRef]
11. Gao, L.; Liu, Y.; Ren, H.; Guerrero, J.M. A DC Microgrid Coordinated Control Strategy Based on Integrator Current-Sharing. *Energies* **2017**, *10*, 1116. [CrossRef]
12. Kumara, J.; Agarwalb, A.; Agarwala, V. A Review on Overall Control of DC Microgrids. *J. Energy Storage* **2019**, *21*, 113–138. [CrossRef]
13. Yazdanian, M.; Mehrizi-Sani, A. Distributed Control Techniques in Microgrids. *IEEE Trans. Smart Grid* **2014**, *5*, 2901–2909. [CrossRef]
14. Logenthiran, T.; Naayagi, R.T.; Woo, W.L.; Phan, V.-T.; Abidi, K. Intelligent Control System for Microgrids Using Multiagent System. *IEEE J. Emerg. Sel. Topics Power Electron.* **2015**, *3*, 1036–1045. [CrossRef]
15. Qi, X.; Luo, Y.H.; Zhang, Y.; Zhou, G.; Wei, Z. Novel Distributed Optimal Control of Battery Energy Storage System in an Islanded Microgrid with Fast Frequency Recovery. *Energies* **2018**, *11*, 1955. [CrossRef]
16. Dehkordi, N.M.; Sadati, N.; Hamzeh, M. Distributed Robust Finite-Time Secondary Voltage and Frequency Control of Islanded Microgrids. *IEEE Trans. Power Syst.* **2017**, *32*, 3648–3659. [CrossRef]
17. Dou, C.; Zhang, Z.; Yue, D.; Zheng, Y. MAS-Based Hierarchical Distributed Coordinate Control Strategy of Virtual Power Source Voltage in Low-Voltage Microgrid. *IEEE Access* **2017**, *5*, 11381–11390. [CrossRef]
18. Lai, J.; Lu, X.; Li, X.; Tang, R.-L. Distributed Multiagent-Oriented Average Control for Voltage Restoration and Reactive Power Sharing of Autonomous Microgrids. *IEEE Access* **2018**, *6*, 25551–25561. [CrossRef]
19. Morstyn, T.; Hredzak, B.; Agelidis, V.G. Communication Delay Robustness for Multi-Agent State of Charge Balancing Between Distributed AC Microgrid Storage Systems. In Proceedings of the 2015 IEEE Conference on Control Applications (CCA), Sydney, Australia, 21–23 September 2015; pp. 181–186.

20. Chen, F.; Chen, M.; Li, Q.; Meng, K.; Guerrero, J.M.; Abbott, D. Multiagent-Based Reactive Power Sharing and Control Model for Islanded Microgrids. *IEEE Trans. Sustain. Energy* **2016**, *7*, 1232–1244. [CrossRef]
21. Li, Q.; Chen, F.; Chen, M.; Guerrero, J.M.; Abbott, D. Agent-Based Decentralized Control Method for Islanded Microgrids. *IEEE Trans. Smart Grid* **2016**, *7*, 637–649. [CrossRef]
22. Dou, C.; Yue, D.; Zhang, Z.; Ma, K. MAS-Based Distributed Cooperative Control for DC Microgrid Through Switching Topology Communication Network with Time-Varying Delays. *IEEE Syst. J.* **2019**, *13*, 615–624. [CrossRef]
23. Lai, J.; Zhou, H.; Hu, W.; Lu, X.; Zhong, L. Synchronization of Hybrid Microgrids with Communication Latency. *Math. Probl. Eng.* **2015**, *2015*, 1–10. [CrossRef]
24. Dou, C.; Yue, D.; Zhang, Z.; Guerrero, J.M. Hierarchical Delay-Dependent Distributed Coordinated Control for DC Ring-Bus Microgrids. *IEEE Access* **2017**, *5*, 10130–10140. [CrossRef]
25. Dou, C.; Yue, D.; Guerrero, J.M.; Xie, X.; Hu, S. Multiagent System-Based Distributed Coordinated Control for Radial DC Microgrid Considering Transmission Time Delays. *IEEE Trans. Smart Grid* **2016**, *8*, 2370–2381. [CrossRef]
26. Zhang, R.; Hredzak, B. Distributed Finite-Time Multiagent Control for DC Microgrids with Time Delays. *IEEE Trans. Smart Grid* **2018**, *10*, 2692–2701. [CrossRef]
27. Zhang, Z.; Dou, C.; Yue, D.; Zhang, B.; Li, F. Neighbor-Prediction-Based Networked Hierarchical Control in Islanded Microgrids. *Int. J. Electr. Power Energy Syst.* **2019**, *104*, 734–743. [CrossRef]
28. Gaeini, N.; Amani, A.M.; Jalili, M.; Yu, X. Cooperative Secondary Frequency Control of Distributed Generation: The Role of Data Communication Network Topology. *Int. J. Electr. Power Energy Syst.* **2017**, *92*, 221–229. [CrossRef]
29. IEEE. *IEEE Standard for Interconnection and Interoperability of Distributed Energy Resources with Associated Electric Power Systems Interfaces, IEEE Std.1547*; IEEE: New York, NY, USA, 2018.
30. Tran, T.V.; Yoon, S.-J.; Kim, K.-H. An LQR-Based Controller Design for an LCL-Filtered Grid-Connected Inverter in Discrete-Time State-Space under Distorted Grid Environment. *Energies* **2018**, *11*, 2062. [CrossRef]
31. Wu, X.; Panda, S.K.; Xu, J. Development of a New Mathematical Model of Three Phase PWM Boost Rectifier Under Unbalanced Supply Voltage Operating Conditions. In Proceedings of the 2006 37th IEEE Power Electronics Specialists Conference, Jeju, Korea, 18–22 June 2006; pp. 1–7.
32. Hegazy, O.; Mierlo, J.V.; Lataire, P. Control and Analysis of an Integrated Bidirectional DC/AC and DC/DC Converters for Plug-in Hybrid Electric Vehicle Applications. *J. Power Electron.* **2011**, *11*, 408–417. [CrossRef]

Article

Optimal Placement of TCSC for Congestion Management and Power Loss Reduction Using Multi-Objective Genetic Algorithm

Thang Trung Nguyen [1] and Fazel Mohammadi [2,*]

[1] Power System Optimization Research Group, Faculty of Electrical and Electronics Engineering, Ton Duc Thang University, Ho Chi Minh City 700000, Vietnam; nguyentrungthang@tdtu.edu.vn

[2] Electrical and Computer Engineering (ECE) Department, University of Windsor, Windsor, ON N9B 1K3, Canada

* Correspondence: fazel@uwindsor.ca or fazel.mohammadi@ieee.org

Received: 20 February 2020; Accepted: 1 April 2020; Published: 2 April 2020

Abstract: Electricity demand has been growing due to the increase in the world population and higher energy usage per capita as compared to the past. As a result, various methods have been proposed to increase the efficiency of power systems in terms of mitigating congestion and minimizing power losses. Power grids operating limitations result in congestion that specifies the final capacity of the system, which decreases the conventional power capabilities between coverage areas. Flexible AC Transmission Systems (FACTS) can help to decrease flows in heavily loaded lines and lead to lines loadability improvements and cost reduction. In this paper, total power loss reduction and line congestion improvement are assessed by determining the optimal locations and compensation rates of Thyristor-Controlled Series Compensator (TCSC) devices using the Multi-Objective Genetic Algorithm (MOGA). The results of applying the proposed method on the IEEE 30-bus test system confirmed the efficiency of the proposed procedure. In addition, to check the performance, applicability, and effectiveness of the proposed method, different heuristic algorithms, such as the multi-objective Particle Swarm Optimization (PSO) algorithm, Differential Evolution (DE) algorithm, and Mixed-Integer Non-Linear Program (MINLP) technique, are used for comparison. The obtained results show the accuracy and fast convergence of the proposed method over the other heuristic techniques.

Keywords: Congestion Management; FACTS devices; Multi-Objective Genetic Algorithm (MOGA); Power Loss Reduction; Thyristor-Controlled Series Compensator (TCSC)

---

1. Introduction

Electric power consumption growth and an open accessible energy market have caused power systems to operate close to their nominal capacities. Also, the extension and development limits of power grids from things, such as installation issues, operating costs, and environmental concerns, have caused many power systems to operate in overload conditions. In addition, power flow in different parts of the grid is restricted by stability and reliability constraints. Therefore, the growth of line power flow exceeding the allowable limits may cause power systems to collapse by random faults [1,2].

These concepts are investigated and studied through the power flow and congestion management topics [3,4]. Transmission lines congestion is a severe problem in power systems operation. The power grid is called congested when some transmission lines operate outside of allowable limits, and as a result, generators may become inactive [5]. Various methods and equipment such as Flexible Alternative Current Transmission Systems (FACTS) devices are reported to manage the active power flow [6]. FACTS devices control the line power flow without any changes in the grid topology, leading to improved performance, increased power transmission capacity, and reduced power grid congestion. Due to the considerable costs of FACTS devices and the maximum usage of their capabilities, the optimal location of such devices should be determined accurately [7–10].

One of the technical challenges in deregulated power systems is congestion management. In [11], the operation of a Thyristor-Controlled Series Compensator (TCSC) for the optimization of transmission lines and transmission line congestion is studied, by developing an algorithm to optimize the performance index for contingency analysis and the location and control of TCSC. The optimal placement of TCSC for improving power transmission efficiency and steady-state stability limits, and maintaining the voltage stability of power systems, is reported in [12]. A method to deal with congestion management by controlling the DC power flow using TCSC is also reported in [13]. In [14], a TCSC is utilized to improve transient stability and congestion management in power systems. In [15], an approach to optimize the location and size of TCSC and, accordingly, reduce the congestion in power systems is investigated. In [16], an approach to find the optimal location and size of a TCSC for congestion management and for enhancing the power transfer capability of power transmission lines, by considering a variable reactance model of the TCSC at the steady-state condition, is investigated. In [17], TCSC location is formulated as the Mixed-Integer Non-Linear Program (MINLP), and an approach for the optimal location and size of TCSC is proposed. In [18], TCSC placement is considered to improve the power transmission line loading parameter, reduce power losses, and improve the voltage stability of power systems. Additionally, the obtained results are compared with allocating the Static VAR Compensator (SVC) for congestion management. The optimal placement and size of TCSC in power systems to reduce the risks in power grid operation is discussed in [19]. The optimal allocation of TCSCs for congestion management is also studied in [20–23] for different applications.

Moreover, total power loss reduction is another important criterion, which should be taken into consideration along with congestion management. In [24], the aim is to reduce power system losses, including switching losses, through economic TCSC installation. Considering short-circuit level and power loss reduction as the two objective functions of the Particle Swarm Optimization (PSO) algorithm, the allocation of TCSC is studied in [25]. In [26], the objective is to provide adequate compensation to reduce the power system losses and improve the voltage profile by finding the optimal location of FACTS devices, including TCSC. As in [26], the optimal allocation of TCSC by checking the sensitivity indices for power loss reduction and voltage profile improvement is investigated in [27].

Congestion management is a systematic approach to schedule and balance the generation and load levels, considering transmission line constraints. Therefore, congestion management in power systems should be continuously investigated due to the fact that changing the generation and load levels can change the location of TCSC. As a result, congestion management requires a fast convergence technique that can obtain optimal solutions. Hence, having a long convergence criterion (number of iterations or computation time) is one of the main drawbacks of other research studies. In addition, obtaining a locally optimal solution instead of a globally optimal one is another issue in this regard. Neglecting non-linear relationships between the parameters of power systems is another issue in other research studies.

To address the above-mentioned issues, this paper aims to optimally allocate TCSCs and their susceptance values, considering power loss reduction, congestion management, and the determination of the power lines compensation rates. The main contributions of this work are (1) to consider the structure of TCSC and the AC characteristics of power systems, formulate a nonlinear problem, and solve it using a heuristic algorithm, and (2) determine the optimal allocation and rating of TCSC through an optimization procedure. The Jacobian sensitivity approach and AC load flow are used for line congestion evaluations. The Multi-Objective Genetic Algorithm (MOGA) is used as an optimization method to determine the optimal locations and susceptance values of TCSCs. To the best of the authors' knowledge, the consideration of power loss reduction, congestion management, and the determination of the power line compensation rates have never been used to optimally allocate TCSCs and their susceptance values. The proposed method is deployed on the IEEE 30-bus test system, and the results are investigated to illustrate the applicability and effectiveness of the proposed method. In addition, the obtained results are compared with those from different algorithms, such as the multi-objective PSO algorithm, Differential Evolution (DE) algorithm, and MINLP technique. The obtained results show the superiorities of the proposed method, in terms of fast convergence and high accuracy, over the other heuristic methods.

The rest of this paper is organized as follows. Section 2 presents the structure and model of TCSCs. The calculations of the power congestion indices of transmission lines are presented in Section 3. Section 4 describes the proposed Multi-Objective Genetic Algorithm (MOGA) and variable coding procedures. The simulation results and discussions are presented in Section 5. Finally, the conclusions are summarized in Section 6.

2. Structure and Model of the TCSC

Series capacitors have been used for many years to improve the stability and loadability of the power transmission grids. The basic principle of their operations is to compensate for the power lines' inductive voltage drop by applying a capacitive voltage and reduce the impact of power transmission line reactance, which can enhance line loadability. One of these series compensators is the TCSC. In the TCSC configuration, a shunt Thyristor Controlled Reactor (TCR) (the set of L and back-to-back connected thyristors, T_1 and T_2) is used in parallel with some parts of a capacitor bank (C). This combination allows the TCSC to provide a reactive component with continuous changes during the thyristors' conduction period. Figure 1 shows a single-phase model of a TCSC connected between bus i and bus j.

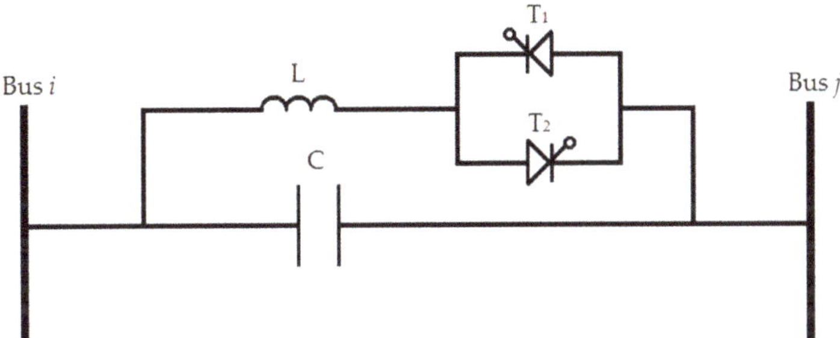

Figure 1. A model of a Thyristor-Controlled Series Compensator (TCSC).

Figure 2 shows the π equivalent parameters of a transmission line when $X >> R$. $V_i\angle\delta_i$ and $V_j\angle\delta_j$ are the complex voltage forms of buses i and j, respectively. In Figure 2, $Y_{ij} = \frac{1}{Z_{ij}} = G_{ij} + jB_{ij}$ is the admittance of the transmission line between buses i and j. G_{ij} and B_{ij} are the conductance and susceptance of the transmission line between buses i and j, respectively. In addition, B_{sh} represents the shunt susceptance of the transmission line.

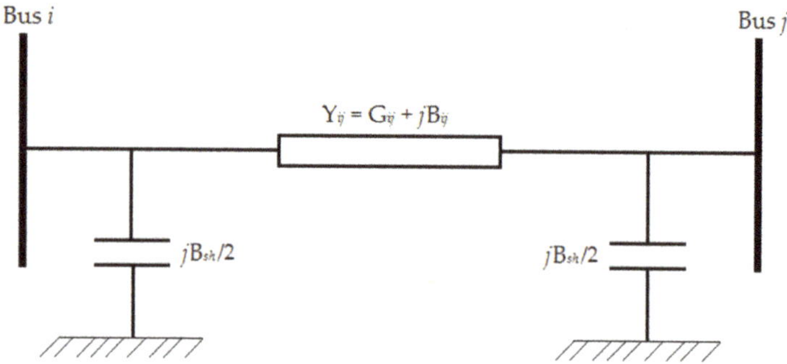

Figure 2. The π equivalent circuit of a transmission line.

The active and reactive power flow equations between bus i and bus j (P_{ij} and Q_{ij}) can be determined as follows [28–33]:

$$P_{ij} = V_i^2 G_{ij} - V_i V_j \left(G_{ij} \cos \delta_{ij} + B_{ij} \sin \delta_{ij} \right) \tag{1}$$

$$Q_{ij} = -V_i{}^2 \left(B_{ij} + B_{sh} \right) - V_i V_j \left(G_{ij} \sin \delta_{ij} - B_{ij} \cos \delta_{ij} \right) \tag{2}$$

where $\delta_{ij} = \delta_i - \delta_j$ is the phase angle difference between the voltage at bus i and bus j.

The transmission line model after the installation of the TCSC between bus i and bus j is shown in Figure 3. In Figure 3, $z_{ij} = r_{ij} + jx_{ij}$ is the impedance of the transmission line between buses i and j. r_{ij} and x_{ij} are the resistance and reactance of the transmission line between buses i and j, respectively. At the steady-state, the TCSC can be considered as a static reactance of $-jx_c$.

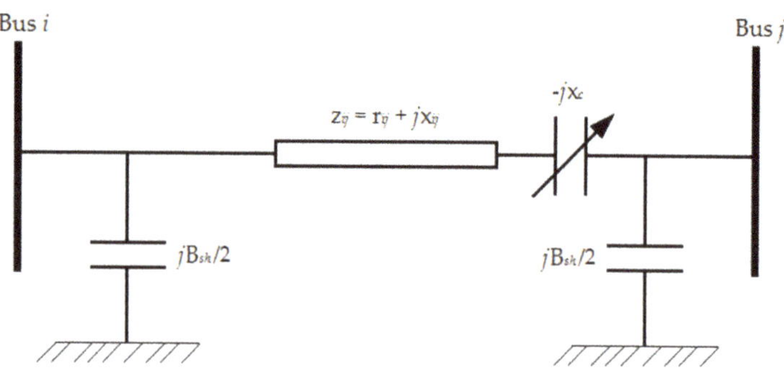

Figure 3. The π equivalent circuit of a transmission line after TSCS installation.

Equations (3) to (6) show the active and reactive power flow equations in the presence of a TCSC between bus i and bus j (P_{ij}^c and Q_{ij}^c) and vice versa (P_{ji}^c and Q_{ji}^c).

$$P_{ij}^c = V_i^2 G'_{ij} - V_i V_j \left(G'_{ij} \cos \delta_{ij} + B'_{ij} \sin \delta_{ij} \right) \tag{3}$$

$$Q_{ij}^c = -V_i^2 \left(B'_{ij} + B_{sh} \right) - V_i V_j \left(G'_{ij} \sin \delta_{ij} - B'_{ij} \cos \delta_{ij} \right) \tag{4}$$

$$P_{ji}^c = V_j^2 G'_{ij} - V_i V_j \left(G'_{ij} \cos \delta_{ij} - B'_{ij} \sin \delta_{ij} \right) \tag{5}$$

$$Q_{ji}^c = -V_j^2 \left(B'_{ij} + B_{sh} \right) + V_i V_j \left(G'_{ij} \sin \delta_{ij} + B'_{ij} \cos \delta_{ij} \right) \tag{6}$$

Active and reactive power losses (P_l and Q_l) can be determined as follows:

$$P_l = P_{ij} + P_{ji} = G'_{ij}\left(V_i^2 + V_j^2\right) - \left(2V_iV_jG'_{ij}\cos\delta_{ij}\right) \tag{7}$$

$$Q_l = Q_{ij} + Q_{ji} = -\left(V_i^2 + V_j^2\right)\left(B_{ij}' + B_{sh}\right) + \left(2V_iV_jB'_{ij}\cos\delta_{ij}\right) \tag{8}$$

where

$$B'_{ij} = \frac{-\left(x_{ij} - x_c\right)}{r_{ij}^2 + \left(x_{ij} - x_c\right)^2} \tag{9}$$

$$G'_{ij} = \frac{r_{ij}}{r_{ij}^2 + \left(x_{ij} - x_c\right)^2} \tag{10}$$

Any changes in the line flow due to the series capacitance can be represented as a line without series capacitance, with power injected at the receiving and sending ends of the line, as shown in Figure 4. S_{ic} and S_{jc} in Figure 4 are the injection power at bus i and bus j, respectively. Details of the active and reactive injection power equations in the presence of a TCSC are expressed in [1–10,27].

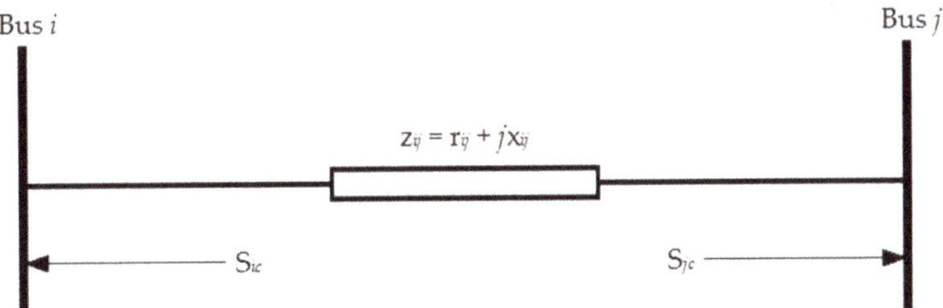

Figure 4. An equivalent injection model of a transmission line after TCSC installation.

3. Power Congestion Index of Transmission Lines

The Transmission Congestion Distribution Factor (TCDF), as proposed in Equation (11), is based on the sensitivity of AC power flow of the lines to changes in the injected power at different buses, as follows [1]:

$$TCDF(i,k) = \frac{\Delta P_{ij}}{\Delta P_i} \tag{11}$$

where $TCDF(i,k)$ represents the variations of active power flow (ΔP_{ij}) in transmission line k between buses i and j due to the changes in the injected power at bus i (ΔP_i). In fact, the $TCDF$ index points to changes in the transmission line power, when the injected power in a bus is changed. There are several methods to determine the TCDF, one of which is presented in this paper.

The power flow equation from bus i to bus j can be expressed as follows [31–33]:

$$P_{ij} = V_iV_jY_{ij}\cos\left(\theta_{ij} + \delta_j - \delta_i\right) - V_i^2Y_{ij}\cos\theta_{ij} \tag{12}$$

where V_i and V_j and δ_i and δ_j are the voltage magnitudes and voltage angles at bus i and bus j, respectively. Y_{ij} and θ_{ij} are the magnitude and phase angle of the admittance of the i-jth element of Y-bus matrix. Using the Taylor series, and ignoring the second- and higher-order terms due to their less impacts, Equations (13) and (14) can be derived as follows:

$$\Delta P_{ij} = \frac{\partial P_{ij}}{\partial \delta_i}\Delta\delta_i + \frac{\partial P_{ij}}{\partial \delta_j}\Delta\delta_j + \frac{\partial P_{ij}}{\partial V_i}\Delta V_i + \frac{\partial P_{ij}}{\partial V_j}\Delta V_j \tag{13}$$

$$\Delta P_{ij} = a_{ij}\Delta\delta_i + b_{ij}\Delta\delta_j + c_{ij}\Delta V_i + d_{ij}\Delta V_j \tag{14}$$

The unknown coefficients in Equation (14) are calculated by deriving Equations (15)–(18).

$$a_{ij} = V_i V_j Y_{ij} \sin\left(\theta_{ij} + \delta_j - \delta_i\right) \quad (15)$$

$$b_{ij} = -V_i V_j Y_{ij} \sin\left(\theta_{ij} + \delta_j - \delta_i\right) \quad (16)$$

$$c_{ij} = V_j Y_{ij} \cos\left(\theta_{ij} + \delta_j - \delta_i\right) - 2V_i Y_{ij} \cos\theta_{ij} \quad (17)$$

$$d_{ij} = V_i Y_{ij} \cos\left(\theta_{ij} + \delta_j - \delta_i\right) \quad (18)$$

Therefore, the Jacobian matrix can be formed as follows:

$$\begin{pmatrix} \Delta P \\ \Delta Q \end{pmatrix} = (j) \begin{pmatrix} \Delta\delta \\ \Delta V \end{pmatrix} = \begin{pmatrix} j_{11} & j_{12} \\ j_{21} & j_{22} \end{pmatrix} \begin{pmatrix} \Delta\delta \\ \Delta V \end{pmatrix} \quad (19)$$

Voltage magnitude changes on the line power flow are ignored due to their negligible impacts. Hence,

$$\Delta P = (j_{11}) \Delta\delta \quad (20)$$

$$\Delta Q = (j_{22}) \Delta V \quad (21)$$

$$\Delta\delta = (j_{11})^{-1} \Delta P = (M) \Delta P \quad (22)$$

Equation (22) can be generally stated as follows:

$$\Delta\delta = \sum_{l=0}^{n} m_{il} \Delta P_l, i = 1, 2, \ldots, n, \qquad i \neq s \quad (23)$$

where n denotes the number of buses, s is slack bus, and m_{il} represents members of matrix M. In accordance with the above-mentioned subject, c_{ij} and d_{ij} are ignored, and Equation (14) can be rewritten as follows:

$$\Delta P_{ij} = a_{ij} \Delta\delta_i + b_{ij} \Delta\delta_j \quad (24)$$

Considering Equations (23) and (24), Equations (25) and (26) can be derived as follows:

$$\Delta P_{ij} = a_{ij} \sum_{l=1}^{n} m_{il} \Delta P_l + b_{ij} \sum_{l=1}^{n} m_{jl} \Delta P_l \quad (25)$$

$$\Delta P_{ij} = TCDF(1,k) \Delta P_1 + TCDF(2,k) \Delta P_2 + \ldots + TCDF(n,k) \Delta P_n \quad (26)$$

where $TCDF(n,k)$ indicates the congestion index of bus n and line k (connection line between bus i and bus j), and it is given as follows:

$$TCDF(n,k) = a_{ij} m_{in} + b_{ij} m_{jn} \quad (27)$$

4. Proposed Multi-Objective Genetic Algorithm (MOGA)

Genetic Algorithm (GA) is a search technique derived from the natural evolutionary mechanism, in which each individual specification of a person is defined by the nature of their chromosomes. There are various applications of the GA, of which multi-objective optimization is one. In fact, the optimization calculations of the GA are performed on variables that form chromosomes using the continuous generation of the population until a predetermined iteration number. On the other hand, the optimization is started with a random generation of the initial population. In the next step, crossover action is performed on chromosomes that are randomly selected. There are several methods for crossover operation, of which the analytical and break methods are two [28–30]. After that, mutation operation is randomly performed on chromosomes. The selection operation is the other most important step in the GA. In this step, by sorting the chromosomes based on their optimum calculated values in the objective function, a defined number of chromosomes, which optimizes the objective function more than the others, will be selected. There are various methods for the chromosome selection, and going to the next step, the Roulette Wheel Algorithm is one of them.

The overall objective function in this problem is composed of the power loss reduction index, congestion improvement indicator (*TCDF*), and TCSC compensation rate. The control (optimization) variables contain the susceptance values of TCSCs ($b_1, \ldots, b_i$) and locations of TCSCs ($B_1, \ldots, B_i$) in power grids, which are discrete numbers between 1 and the maximum line numbers. Figure 5 shows the coding of the problem variables.

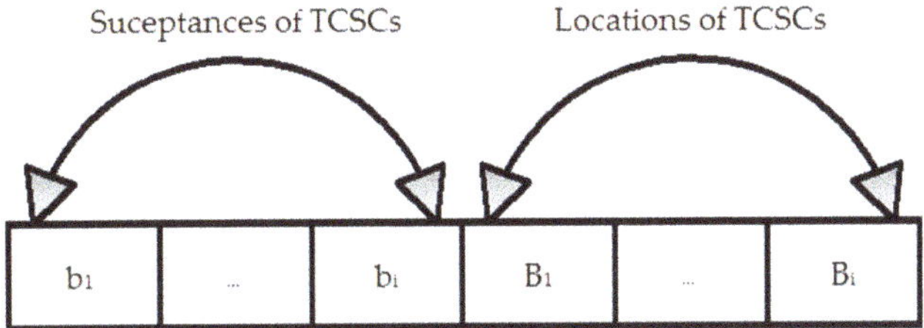

Figure 5. The coding of the objective function variables.

The length of a chromosome is determined according to the number of TCSCs. If one or two TCSCs are utilized, the length of a chromosome is two or four, respectively. Various operational constraints, including the bus voltage, the line maximum power, and also the active and reactive power of generation units are considered in the optimization process [34]. Therefore, the objective function is subjected to the following constraints:

- Bus voltage constraint:

$$V_i^{min} \leq V_i \leq V_i^{max} \tag{28}$$

- Transmission line capacity limitations:

$$S_{L_i} \leq S_{L_i}^{max} \tag{29}$$

- Generator active power limitations:

$$P_{G_i}^{min} \leq P_{G_i} \leq P_{G_i}^{max} \tag{30}$$

- Generator reactive power limitations:

$$Q_{G_i}^{min} \leq Q_{G_i} \leq Q_{G_i}^{max} \tag{31}$$

To handle the inequality constraints, the penalty function is defined as follows:

$$P(X_i) = \begin{cases} \left(X_i - X_i^{max}\right)^2, & X_i > X_i^{max} \\ \left(X_i^{min} - X_i\right)^2, & X_i < X_i^{min} \\ 0, & X_i^{min} < X_i < X_i^{max} \end{cases} \tag{32}$$

where $P(X_i)$ is the penalty function of variable X_i. X_i^{min} and X_i^{max} are the lower and upper limits of variable X_i.

The two objective functions (f_1 and f_2) are presented as follows:

$$f_1 = \sum_{i=1}^{N_L} P_{loss_i} \tag{33}$$

$$f_2 = \frac{1}{N_L}\sum_{i=1}^{N_L}\frac{S_{L_i}}{S_{L_i}^{max}} \times 100 \tag{34}$$

where P_{loss_i} is the active power loss in line i, N_L is the total number of transmission lines, S_{L_i} is the aperient power of line i, and $S_{L_i}^{max}$ denotes the maximum allowable aperient power of line i. It should be noted that in this paper, the average percentage of loadability of the lines is considered as an objective function for congestion minimization.

In addition, the third objective function (f_3) is defined as follows:

$$f_3 = X_L = X_{ij} + X_{TCSC} \tag{35}$$

where X_L is the reactance of the transmission line, X_{ij} is the reactance of the transmission line before compensation, and X_{TCSC} is the added reactance to the transmission line after installation of the TCSC. Also, $X_{TCSC} = r_{TCSC} \times X_{ij}$, where r_{TCSC} is the compensation coefficient and has a value between −0.7 and 0.2.

As a result, the overall objective function can be written as follows:

$$\min F = \sum_{i=1}^{M} w_i f_i \tag{36}$$

where M shows the number of objectives. Also, w_i is the weight factor associated with the ith objective function. It should be noted that $\sum_{i=1}^{K} w_i = 1$. Lastly, f_i is the ith objective function (normalized).

5. Simulation Results and Discussions

5.1. Simulation Results

The proposed method is implemented and evaluated on the IEEE 30-bus test system [35], as shown in Figure 6, for determining the optimal locations of one and two TCSCs. It should be noted that the case-study has 41 transmission lines. MATPOWER is used for power flow analysis [36]. The simulations were accomplished in the MATLAB software using a laptop with the Intel Core i7-8550U processor at 1.80 GHz clock speed and 12 GB of RAM.

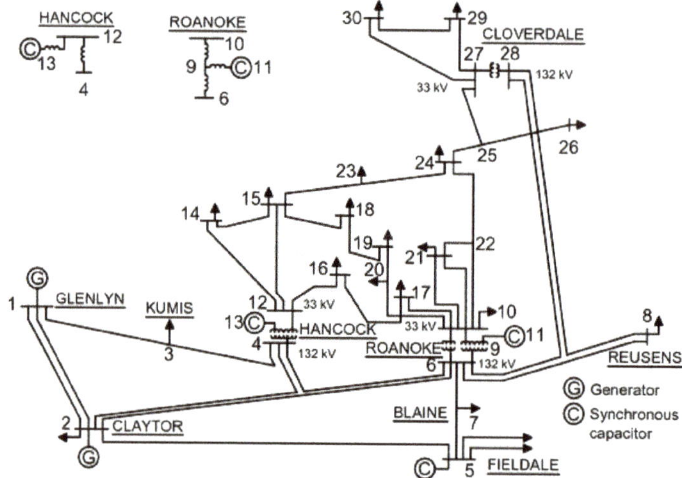

Figure 6. The single-line diagram of the IEEE 30-bus test system.

5.1.1. Installation of One TCSC

In the first scenario, the optimal location of one TCSC and the compensation rate of the corresponding line for power loss reduction are calculated using the MOGA. After 600 iterations, line 36 (the connection line between bus 27 and bus 28) with 59.4% compensation rate of the line reactance is selected as the best location for TCSC installation. Figure 7 demonstrates the voltage improvement at all buses and the voltage drop reduction after the installation of one TCSC.

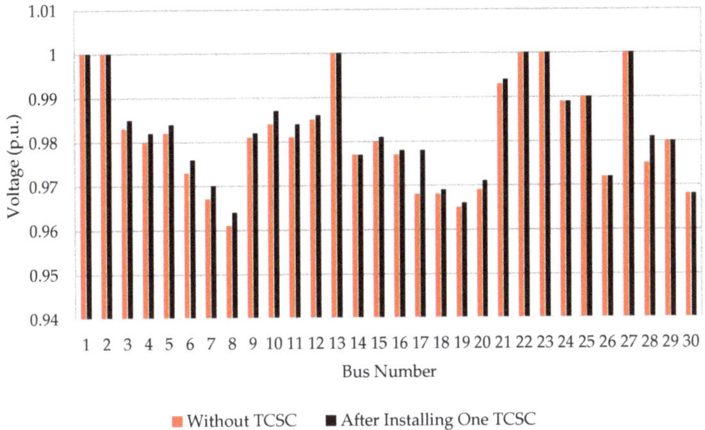

Figure 7. The voltage at different buses before and after the installation of one TCSC.

5.1.2. Installation of Two TCSCs

In the second scenario, the optimization of the objective function is carried out for two TCSCs. After 1000 iterations, lines 36 (the connection line between bus 27 and bus 28) and 16 (the connection line between bus 12 and bus 13), with 59.58% and 56.74% compensation rates of the lines reactance, are chosen as the two optimum TCSC locations. As shown in Figure 8, the total power losses in this case with two TCSCs are expectedly less than the power loss in the last case with one TCSC. It should be noted that the line compensation rate is limited to 60% of the line reactance.

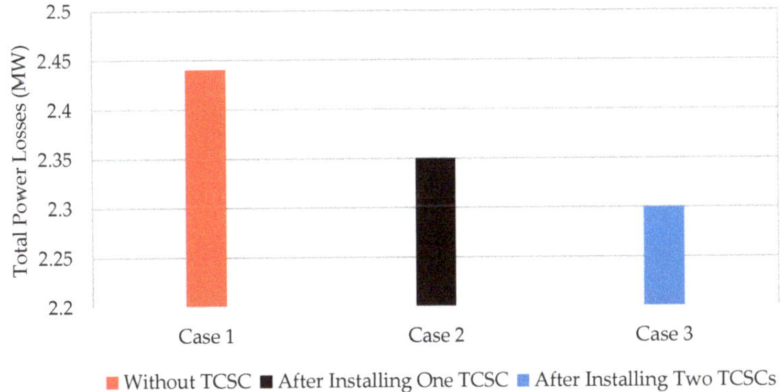

Figure 8. Comparison of the total power losses for different cases.

5.1.3. Optimizing the Congestion Index with One TCSC

In the next scenario, the optimal location and compensation rate of the line, with the aim of optimizing the congestion index using the MOGA, are obtained. For the minimum congestion index (objective function), it is defined that the congestion index should be greater than 2. Therefore, after

600 iterations, line 36 (the connection line between bus 27 and bus 28) is obtained as the optimal location of the TCSC, with 60% compensation rate of the line reactance. Figure 9 shows the congestion index before and after the installation of one TCSC after 600 iterations. In fact, as the congestion index is smaller and closer to zero, the possibility of the overload condition of the lines due to probable variations is less. Therefore, the power flow through the lines can be smoother.

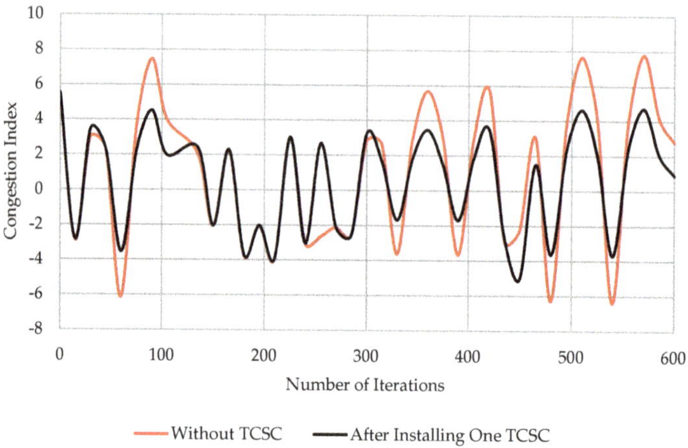

Figure 9. The congestion index before and after installation of one TCSC.

5.1.4. Optimizing the Congestion Index with Two TCSCs

The impact of installing two TCSCs on the system congestion reduction is investigated. After 1000 iterations, it is observed that the best result is obtained by placing TCSCs in lines 36 (the connection line between bus 27 and bus 28) and 12 (the connection line between bus 6 and bus 10) with 59.8% and 55.8% compensation rates of the lines reactance, respectively. Figure 10 shows the obtained congestion index before and after the installation of two TCSCs. It is also observed that the congestion index is reduced in this case.

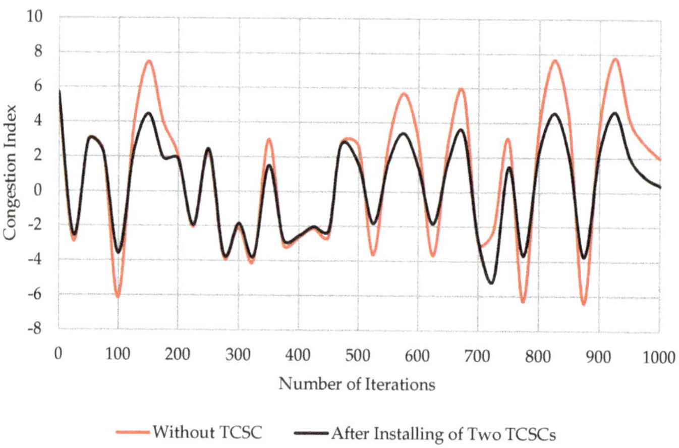

Figure 10. The congestion index before and after installation of two TCSCs.

5.1.5. Comparison of the Impact of TCSC Installation on the Congestion Index

In order to compare the impact of TCSC installation on the congestion index, some of the obtained values for different states of the case-study are given in Table 1. This table shows that the congestion index is reduced slightly after the installation of two TCSCs rather than one. However, some of the congestion indices are increased, the installation of two TCSCs can cause a reduction in the congestion index.

Table 1. Congestion index comparison in three different scenarios.

Without TCSC	With Installing One TCSC	With Installing Two TCSCs
5.6782	5.5209	5.4876
3.4496	3.4496	2.9298
−6.1228	−3.5560	−3.5882
4.5056	1.9668	1.9282
−2.0359	−2.0359	−1.9552
−2.0222	−2.0222	−1.8619
3.0062	3.0062	1.4983
−2.1218	−2.1218	−2.2608
3.2135	3.2135	2.6835
−3.6272	−1.6807	−1.8361
3.0220	1.4799	1.4364
3.4788	1.7925	1.7634
−2.7393	−5.0217	−5.0954
−6.2709	−3.6420	−3.0750
4.1085	2.1043	1.9749
4.6944	2.3643	2.3378

5.2. Performance Evaluation Using Different Heuristic Techniques

To check the performance, applicability, and effectiveness of the proposed method, different heuristic algorithms, such as the multi-objective PSO algorithm, Differential Evolution (DE) algorithm, and MINLP technique are used for comparison, subject to the same conditions (the same population size, same number of iterations, same number of runs, etc.) and on the same machine. The initial population size for each technique is considered as 1000. It should be noted that finite-time and fast convergence is an important capability of any algorithm in practical tests [37–39]. Figures 11–14 show a summary of the comparisons among the proposed method and the multi-objective PSO, DE, and MINLP methods.

Figure 11 shows the voltage improvement at all buses after the installation of one TCSC using different techniques. However, after 600 iterations, line 36 (the connection line between bus 27 and bus 28) is determined as the optimal location of the TCSC using different algorithms, the PSO algorithm shows overall higher compensation rates at all of the busses.

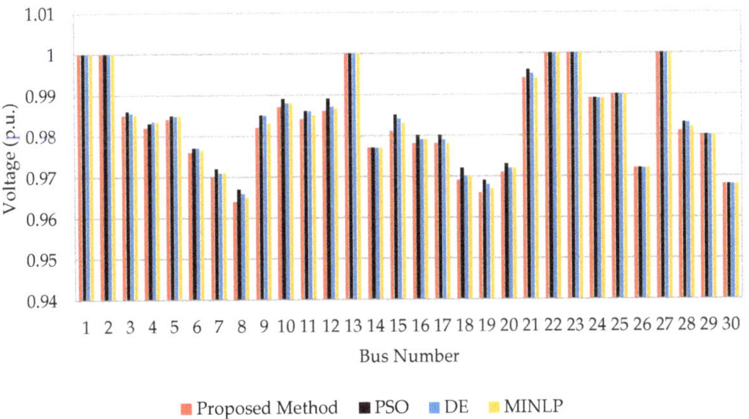

Figure 11. Comparison of the voltage at different buses after the installation of one TCSC using different techniques.

The impact of installing two TCSCs on total power losses using different techniques is shown in Figure 12. As shown in this figure, the MOGA determines the lowest total power losses in all three cases (without TCSC, after installing one TCSC, and after installing two TCSCs).

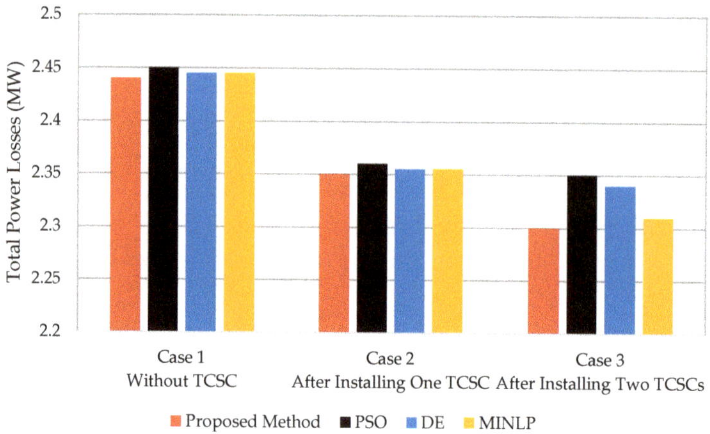

Figure 12. Comparison of the total power losses for different cases using different techniques.

Figure 13 shows the congestion index after the installation of one TCSC using different techniques. As Figure 13 illustrates, the MOGA is highly capable of determining the optimum congestion index after the installation of one TCSC.

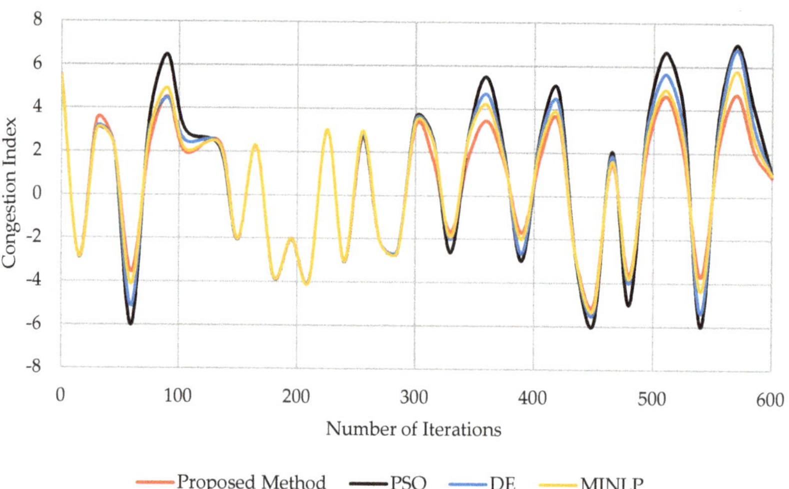

Figure 13. The congestion index after the installation of one TCSC using different techniques.

Figure 14 demonstrates the congestion index after the installation of two TCSCs using different techniques. As shown in this figure, (1) compared to in Figure 14, the congestion index is slightly reduced, and (2) the MOGA is a superior technique to determine the optimum congestion index after the installation of one TCSC.

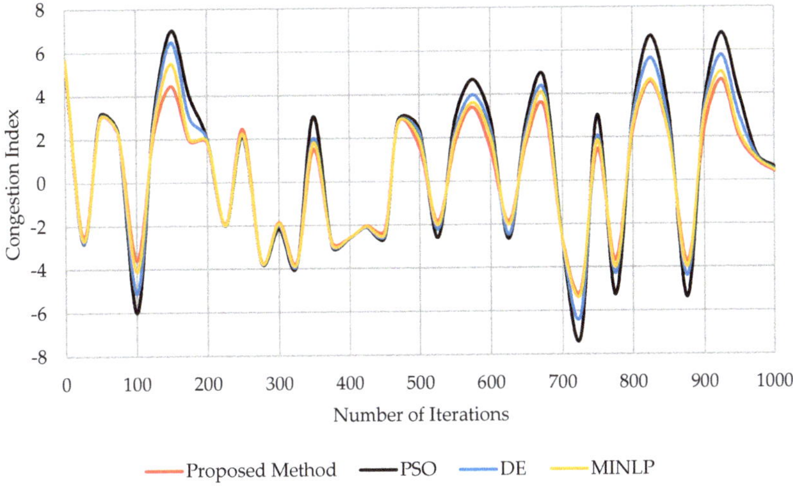

Figure 14. The congestion index after the installation of two TCSCs using different techniques.

6. Conclusions

Congestion management is one of the main challenges of power system optimization. Flexible AC Transmission System (FACTS) devices, such as TCSCs, can be used to manage line congestion with controlling the power flow of the grids. On the other hand, due to the high investment costs of FACTS device installation, determining their optimal locations is very important. In this paper, for the first time, power loss reduction, congestion management, and determination of the power line compensation rates are considered, to optimally allocate TCSCs and their susceptance values. The Jacobian sensitivity approach and AC load flow are used for line congestion evaluations. Then, the Multi-Objective Genetic Algorithm (MOGA) was applied to determine the optimal locations and susceptance values of TCSCs. The obtained results show that the optimal allocation and compensation rate of one TCSC can improve power loss and congestion indices. Additionally, the optimal allocation and compensation rates of two TCSCs show that increasing the number of TCSCs results in a slight power loss reduction and an improvement of the congestion index. Meanwhile, the obtained results are compared with those from different algorithms, such as the multi-objective Particle Swarm Optimization (PSO) algorithm, Differential Evolution (DE) algorithm, and Mixed-Integer Non-Linear Program (MINLP) technique, to evaluate the superiority of the proposed method—in terms of fast convergence and high accuracy—over the other heuristic methods. It is noted that the proposed method is a fast and accurate method that can be used for power system operation and planning studies.

Author Contributions: F.M. was responsible for methodology, collecting resources, data analysis, writing—original draft preparation, supervision, and writing—review and editing. T.-T.N. was responsible for writing—review and editing. All authors have read and agreed to the published version of the manuscript.

Funding: This research received no external funding.

Conflicts of Interest: The authors declare no conflict of interest.

References

1. Besharat, H.; Taher, S.A. Congestion management by determining optimal location of TCSC in deregulated power systems. *Int. J. Electr. Power Energy Syst.* **2008**, *30*. [CrossRef]
2. Rajderkar, V.P.; Chandrakar, V.K. Comparison of series FACTS devices via optimal location in a power system for congestion management. In Proceedings of the Asia-Pacific Power and Energy Engineering Conference, Wuhan, China, 27–31 March 2009.

3. Mandala, M.; Gupta, C.P. Congestion management by optimal placement of facts device. In Proceedings of the Joint International Conference on Power Electronics, Drives and Energy Systems and 2010 Power India, New Delhi, India, 20–23 December 2010.
4. Hosseinipoor, N.; Nabavi, S.M.H. Optimal locating and sizing of TCSC using genetic algorithm for congestion management in deregualted power markets. In Proceedings of the 9th International Conference on Environment and Electrical Engineering, Prague, Czech Republic, 16–19 May 2010.
5. Yousefi, A.; Nguyan, T.T.; Zareipour, H.; Malik, O.P. Congestion management using demand response and FACTS devices. *Int. J. Electr. Power Energy Syst.* **2012**, *37*. [CrossRef]
6. Kumar, A.; Srivastava, S.C.; Singh, S.N. A zonal congestion management approach using AC transmission congestion distribution factors. *Electr. Power Syst. Res.* **2004**. [CrossRef]
7. Mori, H.; Goto, Y. A parallel tabu search based method for determining optimal allocation of FACTS in power systems. In Proceedings of the International Conference on Power System Technology, Perth, Australia, 4–7 December 2000.
8. Singh, S.N.; David, A.K. Congestion management by optimising FACTS device location. In Proceedings of the International Conference on Electric Utility Deregulation and Restructuring and Power Technologies, London, UK, 4–7 April 2000.
9. Gerbex, S.; Cherkaoui, R.; Germond, A.J. Optimal location of multi-type FACTS devices in a power system by means of genetic algorithms. *IEEE Trans. Power Syst.* **2001**, *16*. [CrossRef]
10. Kumar, A.; Sekhar, C. Congestion management with FACTS devices in deregulated electricity markets ensuring loadability limit. *Int. J. Electr. Power Energy Syst.* **2013**. [CrossRef]
11. Vetrivel, K.; Mohan, G.; Rao, K.U.; Bhat, S.H. Optimal placement and control of TCSC for transmission congestion management. In Proceedings of the International Conference on Computation of Power, Energy, Information and Communication (ICCPEIC), Chennai, India, 16–17 April 2014.
12. Choudekar, P.; Sinha, S.K.; Siddiqui, A. Transmission line efficiency improvement and congestion management under critical contingency condition by optimal placement of TCSC. In Proceedings of the 7th India International Conference on Power Electronics (IICPE), Patiala, India, 17–19 November 2016.
13. Surya, R.; Janarthanan, N.; Balamurugan, S. A novel technique for congestion management in transmission system by real power flow control. In Proceedings of the International Conference on Intelligent Computing, Instrumentation and Control Technologies (ICICICT), Kannur, India, 6–7 July 2017.
14. Retnamony, R.; Raglend, I.J. Congestion management is to enhance the transient stability in a deregulated power system using FACTS devices. In Proceedings of the International Conference on Control, Instrumentation, Communication and Computational Technologies (ICCICCT), Kumaracoil, India, 18–19 December 2015.
15. Dawn, S.; Tiwari, P.K.; Gope, S.; Goswami, A.K.; Kumar, P. An active power spot price based approach for congestion management by optimal allocation of TCSC in competitive power market. In Proceedings of the IEEE Region 10 Conference (TENCON), Singapore, Singapore, 22–25 November 2016.
16. Kulkarni, P.P.; Ghawghawe, N.D. Optimal placement and parameter setting of tcsc in power transmission system to increase the power transfer capability. In Proceedings of the International Conference on Energy Systems and Applications, Pune, India, 30 October–1 November 2015.
17. Ziaee, O.; Choobineh, F.F. Optimal location-allocation of TCSC devices on a transmission network. *IEEE Trans. Power Syst.* **2017**, *32*. [CrossRef]
18. Khatavkar, V.; Namjoshi, M.; Dharme, A. Congestion management in deregulated electricity market using FACTS and multi-objective optimization. In Proceedings of the Indian Control Conference (ICC), Hyderabad, India, 4–6 January 2016.
19. Shchetinin, D.; Hug, G. Optimal TCSC allocation in a power system for risk minimization. In Proceedings of the North American Power Symposium (NAPS), Pullman, WA, USA, 7–9 September 2014.
20. Siddiqui, A.S.; Khan, M.T.; Iqbal, F. A novel approach to determine optimal location of TCSC for congestion management. In Proceedings of the 6th IEEE Power India International Conference (PIICON), Delhi, India, 5–7 December 2014.
21. Tang, W.; Pan, W.; Zhang, K.; Liu, C. Study of the location of TCSC for improving ATC considering N-1 security constraints. In Proceedings of the 26th Chinese Control and Decision Conference (CCDC), Changsha, China, 31 May–2 June 2014.

22. Sarwar, M.; Khan, M.T.; Siddiqui, A.S.; Quadri, I.A. An approach to locate TCSC optimally for congestion management in deregulated electricity market. In Proceedings of the 7th India International Conference on Power Electronics (IICPE), Patiala, India, 17–19 November 2016.
23. Sharma, A.; Jain, S. Locating series FACTS devices for managing transmission congestion in deregulated power market. In Proceedings of the 7th India International Conference on Power Electronics (IICPE), Patiala, India, 17–19 November 2016.
24. Gitizadeh, M.; Khalilnezhad, H. Power system loss reduction through TCSC economic installation considering switching loss. In Proceedings of the 45th International Universities Power Engineering Conference (UPEC), Cardiff, Wales, UK, 31 August–3 September 2010.
25. Baghaee, H.R.; Kaviani, A.K.; Mirsalim, M.; Gharehpetian, G.B. Short circuit level and loss reduction by allocating TCSC and UPFC using particle swarm optimization. In Proceedings of the 19th Iranian Conference on Electrical Engineering, Tehran, Iran, 17–19 May 2011.
26. Hridya, K.R.; Mini, V.; Visakhan, R.; Kurian, A.A. Analysis of voltage stability enhancement of a grid and loss reduction using series FACTS controllers. In Proceedings of the International Conference on Power, Instrumentation, Control and Computing (PICC), Thrissur, India, 9–11 December 2015.
27. Manganuri, Y.; Choudekar, P.; Abhishek; Asija, D.; Ruchira. Optimal location of TCSC using sensitivity and stability indices for reduction in losses and improving the voltage profile. In Proceedings of the 1st International Conference on Power Electronics, Intelligent Control and Energy Systems (ICPEICES), Delhi, India, 4–6 July 2016.
28. Moslemi, R.; Shayanfar, H.A. Optimal location for series FACTS devices to transient stability constrained congestion management. In Proceedings of the 10th International Conference on Environment and Electrical Engineering, Rome, Italy, 8–11 May 2011.
29. Reddy, S.S.; Kumari, M.S.; Sydulu, M. Congestion management in deregulated power system by optimal choice and allocation of FACTS controllers using multi-objective genetic algorithm. *IEEE PES* **2010**. [CrossRef]
30. Abou-Ghazala, A. Optimal capacitor placement in distribution systems feeding nonlinear loads. In Proceedings of the IEEE Bologna Power Tech Conference Proceedings, Bologna, Italy, 23–26 June 2003.
31. Mohammadi, F.; Zheng, C. Stability analysis of electric power system. In Proceedings of the 4th National Conference on Technology in Electrical and Computer Engineering, Tehran, Iran, 27 December 2018.
32. Mohammadi, F.; Nazri, G.-A.; Saif, M. A bidirectional power charging control strategy for plug-in hybrid electric vehicles. *Sustainability* **2019**, *11*, 4317. [CrossRef]
33. Mohammadi, F.; Nazri, G.-A.; Saif, M. An improved mixed AC/DC power flow algorithm in hybrid AC/DC grids with MT-HVDC systems. *Appl. Sci.* **2020**, *10*, 297. [CrossRef]
34. Khanabadi, M.; Ghasemi, H. Transmission congestion management through optimal transmission switching. In Proceedings of the IEEE Power and Energy Society General Meeting, Detroit, MI, USA, 24–28 July 2011.
35. Christie, R. *Power Systems Test Case Archive*; UW Power Systems Test Case Archive: Seattle, WA, USA, 1993.
36. Zimmerman, R.D.; Murillo-Sanchez, C.E.; Thomas, R.J. MATPOWER: Steady-state operations, planning and analysis tools for power systems research and education. *IEEE Trans. Power Syst.* **2011**, *26*. [CrossRef]
37. Mohammadi, F.; Nazri, G.-A.; Saif, M. A fast fault detection and identification approach in power distribution systems. In Proceedings of the 5th International Conference on Power Generation Systems and Renewable Energy Technologies (PGSRET 2019), Istanbul, Turkey, 26–27 August 2019.
38. Karimi, H.; Ghasemi, R.; Mohammadi, F. Adaptive neural observer-based nonsingular terminal sliding mode controller design for a class of nonlinear systems. In Proceedings of the 6th International Conference on Control, Instrumentation, and Automation (ICCIA 2019), Kordestan, Iran, 30–31 October 2019.
39. Tavoosi, J.; Mohammadi, F. Design a new intelligent control for a class of nonlinear systems. In Proceedings of the 6th International Conference on Control, Instrumentation, and Automation (ICCIA 2019), Kordestan, Iran, 30–31 October 2019.

MDPI
St. Alban-Anlage 66
4052 Basel
Switzerland
Tel. +41 61 683 77 34
Fax +41 61 302 89 18
www.mdpi.com

Sustainability Editorial Office
E-mail: sustainability@mdpi.com
www.mdpi.com/journal/sustainability

www.ingramcontent.com/pod-product-compliance
Lightning Source LLC
LaVergne TN
LVHW070702170726
843515LV00004B/464

* 9 7 8 3 0 3 9 3 6 1 8 0 9 *